HOW WE
GOT TO NOW

How We Got to Now

SIX INNOVATIONS THAT MADE THE MODERN WORLD

Steven Johnson

PARTICULAR BOOKS

an imprint of

PENGUIN BOOKS

PARTICULAR BOOKS

Published by the Penguin Group
Penguin Books Ltd, 80 Strand, London WC2R ORL, England
Penguin Group (USA) Inc., 375 Hudson Street, New York, New York 10014, USA
Penguin Group (Canada), 90 Eglinton Avenue East, Suite 700, Toronto, Ontario, Canada M4P 2Y3
(a division of Pearson Penguin Canada Inc.)
Penguin Ireland, 25 St Stephen's Green, Dublin 2, Ireland (a division of Penguin Books Ltd)
Penguin Group (Australia), 707 Collins Street, Melbourne, Victoria 3008, Australia
(a division of Pearson Australia Group Pty Ltd)
Penguin Books India Pvt Ltd, 11 Community Centre, Panchsheel Park, New Delhi – 110 017, India
Penguin Group (NZ), 67 Apollo Drive, Rosedale, Auckland 0632, New Zealand
(a division of Pearson New Zealand Ltd)
Penguin Books (South Africa) (Pty) Ltd, Block D, Rosebank Office Park,
181 Jan Smuts Avenue, Parktown North, Gauteng 2193, South Africa

Penguin Books Ltd, Registered Offices: 80 Strand, London WC2R ORL, England

www.penguin.com

First published in the United States of America by Penguin Group (USA) LLC
First published in Great Britain by Particular Books 2014
001

Set in 13.5/18 pt Garamond MT Std
Typeset by Jouve (UK), Milton Keynes
Printed in Great Britain by Clays Ltd, St Ives plc

A CIP catalogue record for this book is available from the British Library

HARDBACK ISBN: 978–1–846–14853–8
TRADE PAPERBACK ISBN: 978–1–846–14860–6

www.greenpenguin.co.uk

For Jane, who no doubt expected a three-volume treatise on nineteenth-century whaling

Contents

List of Illustrations

INTRODUCTION

Robot Historians and the Hummingbird's Wing

A little more than two decades ago, the Mexican-American artist and philosopher Manuel De Landa published a strange and wonderful book called *War in the Age of Intelligent Machines.* The book was, technically speaking, a history of military technology, but it had nothing in common with what you might naturally expect from the genre. Instead of heroic accounts of submarine engineering written by some Naval Academy professor, De Landa's book wove chaos theory, evolutionary biology, and French post-structuralist philosophy into histories of the conoidal bullet, radar, and other military innovations. I remember reading it as a grad student in my early twenties and thinking that it was one of those books that seemed completely *sui generis*, as though De Landa had arrived on Earth from some other intellectual planet. It seemed mesmerizing and deeply confusing at the same time.

De Landa began the book with a brilliant interpretative twist. Imagine, he suggested, a work of history written sometime in the future by some form of artificial intelligence, mapping out the history of the preceding millennium. 'We could imagine,' De Landa argued, 'that such a robot historian would write a different kind of history than would its human counterpart.' Events that loom large in human accounts – the European conquest of the Americas, the fall of the Roman Empire, the Magna Carta – would be footnotes from the robot's perspective. Other events that seem marginal to traditional history – the toy automatons that pretended to play chess in the eighteenth century, the Jacquard loom that inspired the punch cards of early computing – would be watershed moments to the robot historian, turning points that trace a direct line to the present. 'While a human historian might try to understand the way people assembled clockworks, motors and other physical contraptions,' De Landa explained, 'a robot historian would likely place a stronger emphasis on the way these machines affected human evolution. The robot would stress the fact that when clockworks once represented the dominant technology on the planet, people imagined the world around them as a similar system of cogs and wheels.'

There are no intelligent robots in this book, alas. The innovations here belong to everyday life, not science fiction: lightbulbs, sound recordings, air-conditioning, a glass of clean tap water, a wristwatch, a glass lens. But I have tried to tell the story of these innovations from something like the perspective of De Landa's robot historian. If the lightbulb could write a history of the past three hundred years, it too would look very different. We would see how much of our past was bound up in the pursuit of artificial light, how much ingenuity and struggle went into the battle against darkness, and how the inventions we came up with triggered changes that, at first glance, would seem to have nothing to do with lightbulbs.

This is a history worth telling, in part, because it allows us to see a world we generally take for granted with fresh eyes. Most of us in the developed world don't pause to think how amazing it is that we drink water from a tap and never once worry about dying forty-eight hours later from cholera. Thanks to air-conditioning, many of us live comfortably in climates that would have been intolerable just fifty years ago. Our lives are surrounded and supported by a whole class of objects that are enchanted with the ideas and creativity of thousands of people who came before us: inventors and hobbyists and reformers who steadily hacked away at the problem of making artificial light or clean drinking water so that we can enjoy those luxuries today without a second thought, without even thinking of them as luxuries in the first place. As the robot historians would no doubt remind us, we are indebted to those people every bit as much as, if not more than, we are to the kings and conquerors and magnates of traditional history.

But the other reason to write this kind of history is that these innovations have set in motion a much wider array of changes in society than you might reasonably expect. Innovations usually begin life with an attempt to solve a specific problem, but once they get into circulation, they end up triggering other changes that would have been extremely difficult to predict. This is a pattern of change that appears constantly in evolutionary history. Think of the act of pollination: sometime during the Cretaceous age, flowers began to evolve colors and scents that signaled the presence of pollen to insects, who simultaneously evolved complex equipment to extract the pollen and, inadvertently, fertilize other flowers with pollen. Over time, the flowers supplemented the pollen with even more energy-rich nectar to lure the insects into the rituals of pollination. Bees and other insects evolved the sensory tools to see and be drawn to flowers, just as the flowers evolved the properties that attract bees. This is a different kind of survival of the fittest, not the usual zero-sum competitive story that we often hear in watered-down versions of Darwinism, but something more symbiotic: the insects and flowers succeed because they, physically, fit well with each other. (The technical term for this is coevolution.) The importance of this relationship was not lost on Charles Darwin, who followed up the publication of *On the Origin of Species* with an entire book on orchid pollination.

These coevolutionary interactions often lead to transformations in organisms that would seem to have no immediate connection to the original species. The symbiosis between flowering plants and insects that led to the production of nectar ultimately created an opportunity for much larger organisms – the hummingbirds – to extract nectar from plants, though to do that they evolved an extremely unusual form of flight mechanics that enables them to hover alongside the flower in a way that few birds can even come close to doing. Insects can stabilize themselves midflight because they have fundamental flexibility to their anatomy that vertebrates lack. Yet despite the restrictions placed on them by their skeletal structure, hummingbirds evolved a novel way of rotating their wings, giving power to the upstroke as well as the downstroke, enabling them to float midair while extracting nectar from a flower. These are the strange leaps that evolution makes constantly: the sexual reproduction strategies of plants end up shaping the design of a hummingbird's wings. Had there been naturalists around to observe the insects first evolving pollination behavior alongside the flowering plants, they would have logically assumed that this strange new ritual had nothing to do with avian life. And yet it ended up precipitating one of the most astonishing physical transformations in the evolutionary history of birds.

The history of ideas and innovation unfolds the same way. Johannes Gutenberg's printing press created a surge in demand for spectacles, as the new practice of reading made Europeans across the continent suddenly realize that they were farsighted; the market demand for spectacles encouraged a growing number of people to produce and experiment with lenses, which led to the invention of the microscope, which shortly thereafter enabled us to perceive that our bodies were made up of microscopic cells. You wouldn't think that printing technology would have anything to do with the expansion of our vision down to the cellular scale, just as you wouldn't have thought that the evolution of pollen would alter the design of a hummingbird's wing. But that is the way change happens.

This may sound, at first blush, like a variation on the famous 'butterfly effect' from chaos theory, where the flap of a butterfly's wing in California ends up triggering a hurricane in the mid-Atlantic. But in fact, the two are fundamentally different. The extraordinary (and unsettling) property of the butterfly effect is that it involves a virtually unknowable chain of causality; you can't map the link between the air molecules bouncing around the butterfly and the storm system brewing in the Atlantic. They may be connected, because everything is connected on some level, but it is beyond our capacity to parse those connections or, even harder, to predict them in advance. But something very different is at work with the flower and the hummingbird: while they are very different organisms, with very different needs and aptitudes, not to mention basic biological systems, the flower clearly influences the hummingbird's physiognomy in direct, intelligible ways.

This book is then partially about these strange chains of influence, the 'hummingbird effect.' An innovation, or cluster of innovations, in one field ends up triggering changes that seem to belong to a different domain altogether. Hummingbird effects come in a variety of forms. Some are intuitive enough: orders-of-magnitude increases in the sharing of energy or information tend to set in motion a chaotic wave of change that easily surges over intellectual and social boundaries. (Just look at the story of the Internet over the past thirty years.) But other hummingbird effects are more subtle; they leave behind less conspicuous causal fingerprints. Breakthroughs in our ability to measure a phenomenon – time, temperature, mass – often open up new opportunities that seem at first blush to be unrelated. (The pendulum clock helped enable the factory towns of the industrial revolution.) Sometimes, as in the story of Gutenberg and the lens, a new innovation creates a liability or weakness in our natural toolkit, that sets us out in a new direction, generating new tools to fix a 'problem' that was itself a kind of invention. Sometimes new tools reduce natural barriers and limits to human growth, the way the invention of air-conditioning enabled humans to colonize the hotspots of the planet at a scale that would have startled our ancestors just three generations ago. Sometimes the new tools influence us metaphorically, as in the robot historian's connection between the clock and the mechanistic view of early physics, the universe imagined as a system of 'cogs and wheels.'

Observing hummingbird effects in history makes it clear that social transformations are not always the direct result of human agency and decision-making. Sometimes change comes about through the actions of political leaders or inventors or protest movements, who deliberately bring about some kind of new reality through their conscious planning. (We have an integrated national highway system in the United States in large part because our political leaders decided to pass the Federal-Aid Highway Act of 1956.) But in other cases, the ideas and innovations seem to have a life of their own, engendering changes in society that were not part of their creators' vision. The inventors of air-conditioning were not trying to redraw the political map of America when they set about to cool down living rooms and office buildings, but, as we will see, the technology they unleashed on the world enabled dramatic changes in American settlement patterns, which in turn transformed the occupants of Congress and the White House.

I have resisted the understandable temptation to assess these changes with some kind of value judgment. Certainly this book is a celebration of our ingenuity, but just because an innovation happens, that doesn't mean there aren't, in the end, mixed consequences as it ripples through society. Most ideas that get 'selected' by culture are demonstrably improvements in terms of local objectives: the cases where we have chosen an inferior technology or scientific principle over a more productive or accurate one are the exceptions that prove the rule. And even when we do briefly choose the inferior VHS over Betamax, before long we have DVDs that outperform either option. So when you look at the arc of history from that perspective, it does trend toward better tools, better energy sources, better ways to transmit information.

The problem lies with the externalities and unintended consequences. When Google launched its original search tool in 1999, it was a momentous improvement over any previous technique for exploring the Web's vast archive. That was cause for celebration on almost every level: Google made the entire Web more useful, for free. But then Google started selling advertisements tied into the search requests it received, and within a few years, the efficiency of the searches (along with a few other online services like Craigslist) had hollowed out the advertising base of local newspapers around the United States. Almost no one saw that coming, not even the Google founders. You can make the argument – as it happens, I would probably make the argument – that the trade-off was worth it, and that the challenge from Google will ultimately unleash better forms of journalism, built around the unique opportunities of the Web instead of the printing press. But certainly there is a case to be made that the rise of Web advertising has been, all told, a negative development for the essential public resource of newspaper journalism. The same debate rages over just about every technological advance: Cars moved us more efficiently through space than did horses, but were they worth the cost to the environment or the walkable city? Air-conditioning allowed us to live in deserts, but at what cost to our water supplies?

This book is resolutely agnostic on these questions of value. Figuring out whether we think the change is better for us in the long run is not the same as figuring out how the change came about in the first place. Both kinds of figuring are essential if we are to make sense of history and to map our path into the future. We need to be able to understand how innovation happens in society; we need to be able to predict and understand, as best as we can, the hummingbird effects that will transform other fields after each innovation takes root. And at the same time we need a value system to decide which strains to encourage and which benefits aren't worth the tangential costs. I have tried to spell out the full range of consequences with the innovations surveyed in this book, the good and the bad. The vacuum tube helped bring jazz to a mass audience, and it also helped amplify the Nuremberg rallies. How you ultimately feel about these transformations – Are we ultimately better off thanks to the invention of the vacuum tube? – will depend on your own belief systems about politics and social change.

I should mention one additional element of the book's focus: The 'we' in this book, and in its title, is largely the 'we' of North Americans and Europeans. The story of how China or Brazil got to now would be a different one, and every bit as interesting. But the European/North American story, while finite in its scope, is nonetheless of wider relevance because certain critical experiences – the rise of the scientific method, industrialization – happened in Europe *first*, and have now spread across the world. (*Why* they happened in Europe first is of course one of the most interesting questions of all, but it's not one this book tries to answer.) Those enchanted objects of everyday life – those lightbulbs and lenses and audio recordings – are now a part of life just about everywhere on the planet; telling the story of the past thousand years from their perspective should be of interest no matter where you happen to live. New innovations are shaped by geopolitical history; they cluster in cities and trading hubs. But in the long run, they don't have a lot of patience for borders and national identities, never more so than now in our connected world.

I have tried to adhere to this focus because, within these boundaries, the history I've written here is in other respects as expansive as possible. Telling the story of our ability to capture and transmit the human voice, for instance, is not just a story about a few brilliant inventors, the Edisons and Bells whose names every schoolchild has already memorized. It's also a story about eighteenth-century anatomical drawings of the human ear, the sinking of the *Titanic*, the civil rights movement, and the strange acoustic properties of a broken vacuum tube. This is an approach I have elsewhere called 'long zoom' history: the attempt to explain historical change by simultaneously examining multiple scales of experience – from the vibrations of sound waves on the eardrum all the way out to mass political movements. It may be more intuitive to keep historical narratives on the scale of individuals or nations, but on some fundamental level, it is not accurate to remain between those boundaries. History happens on the level of atoms, the level of planetary climate change, and all the levels in between. If we are trying to get the story right, we need an interpretative approach that can do justice to all those different levels.

The physicist Richard Feynman once described the relationship between aesthetics and science in a similar vein:

> I have a friend who's an artist and has sometimes taken a view which I don't agree with very well. He'll hold up a flower and say 'Look how beautiful it is,' and I'll agree. Then he says 'I as an artist can see how beautiful this is but you as a scientist take this all apart and it becomes a dull thing,' and I think that he's kind of nutty. First of all, the beauty that he sees is available to other people and to me too, I believe. Although I may not be quite as refined aesthetically as he is . . . I can appreciate the beauty of a flower. At the same time, I see much more about the flower than he sees. I could imagine the cells in there, the complicated actions inside, which also have a beauty. I mean it's not just beauty at this dimension, at one centimeter; there's also beauty at smaller dimensions, the inner structure, also the processes. The fact that the colors in the flower evolved in order to attract insects to pollinate it is interesting; it means that insects can see the color. It adds a question: does this aesthetic sense also exist in the lower forms? Why is it aesthetic? All kinds of interesting questions which shows that a science knowledge only adds to the excitement, the mystery and the awe of a flower. It only adds. I don't understand how it subtracts.

There is something undeniably appealing about the story of a great inventor or scientist – Galileo and his telescope, for instance – working his or her way toward a transformative idea. But there is another, deeper story that can be told as well: how the ability to make lenses also depended on the unique quantum mechanical properties of silicon dioxide and on the fall of Constantinople. Telling the story from that long-zoom perspective doesn't subtract from the traditional account focused on Galileo's genius. It only adds.

Marin County, California
February 2014

CHAPTER 1

Glass

Roughly 26 million years ago, something happened over the sands of the Libyan Desert, the bleak, impossibly dry landscape that marks the eastern edge of the Sahara. We don't know exactly what it was, but we do know that it was hot. Grains of silica melted and fused under an intense heat that must have been at least a thousand degrees. The compounds of silicon dioxide they formed have a number of curious chemical traits. Like H_2O, they form crystals in their solid state, and melt into a liquid when heated. But silicon dioxide has a much higher melting point than water; you need temperatures above 500 degrees Fahrenheit instead of 32. But the truly peculiar thing about silicon dioxide is what happens when it cools. Liquid water will happily re-form the crystals of ice if the temperature drops back down again. But silicon dioxide for some reason is incapable of rearranging itself back into the orderly structure of crystal. Instead, it forms a new substance that exists in a strange limbo between solid and liquid, a substance human beings have been obsessed with since the dawn of civilization. When those superheated grains of sand cooled down below their melting point, a vast stretch of the Libyan Desert was coated with a layer of what we now call glass.

About ten thousand years ago, give or take a few millennia, someone traveling through the desert stumbled across a large fragment of this glass. We don't know anything more about that fragment, only that it must have impressed just about everyone who came into contact with it, because it circulated through the markets and social networks of early civilization, until it ended up as a centerpiece of a brooch, carved into the shape of a scarab beetle. It sat there undisturbed for four thousand years, until archeologists unearthed it in 1922 while exploring the tomb of an Egyptian ruler. Against all odds, that small sliver of silicon dioxide had found its way from the Libyan Desert into the burial chamber of Tutankhamun.

Glass first made the transition from ornament to advanced technology during the height of the Roman Empire, when glassmakers figured out ways to make the material sturdier and less cloudy than naturally forming glass like that of King Tut's scarab. Glass windows were built during this period for the first time, laying the groundwork for the shimmering glass towers that now populate city skylines around the world. The visual aesthetics of drinking wine emerged as people consumed it in semitransparent glass vessels and stored it in glass bottles. But, in a way, the early history of glass is relatively predictable: craftsmen figured out how to melt the silica into drinking vessels or windowpanes, exactly the sort of typical uses we instinctively associate with glass today. It wasn't until the next millennium, and the fall of another great empire, that glass became what it is today: one of the most versatile and transformative materials in all of human culture.

The sacking of Constantinople in 1204 was one of those historical quakes that send tremors of influence rippling across the globe. Dynasties fall, armies surge and retreat, the map of the world is redrawn. But the fall of Constantinople also triggered a seemingly minor event, lost in the midst of that vast reorganization of religious and geopolitical dominance and ignored by most historians of the time. A small community of glassmakers from Turkey sailed westward across the Mediterranean and settled in Venice, where they began practicing their trade in the prosperous new city growing out of the marshes on the shores of the Adriatic Sea.

It was one of a thousand migrations set in motion by Constantinople's fall, but looking back over the centuries, it turned out to be one of the most significant. As they settled into the canals and crooked streets of Venice, at that point arguably the most important hub of commercial trade in the world, their skills at blowing glass quickly created a new luxury good for the merchants of the city to sell around the globe. But lucrative as it was, glassmaking was not without its liabilities. The melting point of silicon dioxide required furnaces burning at temperatures near 1,000 degrees, and Venice was a city built almost entirely out of wooden structures. (The classic stone Venetian palaces would not be built for another few centuries.) The glassmakers had brought a new source of wealth to Venice, but they had also brought the less appealing habit of burning down the neighborhood.

In 1291, in an effort to both retain the skills of the glassmakers and protect public safety, the city government sent the glassmakers into exile once again, only this time their journey was a short one – a mile across the Venetian Lagoon to the island of Murano. Unwittingly, the Venetian doges had created an innovation hub: by concentrating the glassmakers on a single island the size of a small city neighborhood, they triggered a surge of creativity, giving birth to an environment that possessed what economists call 'information spillover.' The density of Murano meant that new ideas were quick to flow through the entire population. The glassmakers were in part competitors, but their family lineages were heavily intertwined. There were individual masters in the group that had more talent or expertise than the others, but in general the genius of Murano was a collective affair: something created by sharing as much as by competitive pressures.

By the first years of the next century, Murano had become known as the Isle of Glass, and its ornate vases and other exquisite glassware became status symbols throughout Western Europe. (The glassmakers continue to work their trade today, many of them direct descendants of the original families that emigrated from Turkey.) It was not exactly a model that could be directly replicated in modern times: mayors looking to bring the creative class to their cities probably shouldn't consider forced exile and borders armed with the death penalty. But somehow it worked. After years of trial and error, experimenting with different chemical compositions, the Murano glassmaker Angelo Barovier took seaweed rich in potassium oxide and manganese, burned it to create ash, and then added these ingredients to molten glass. When the mixture cooled, it created an extraordinarily clear type of glass. Struck by its resemblance to the clearest rock crystals of quartz, Barovier called it *cristallo*. This was the birth of modern glass.

While glassmakers such as Barovier were brilliant at making glass transparent, we didn't understand scientifically *why* glass is transparent until the twentieth century. Most materials absorb the energy of light. On a subatomic level, electrons orbiting the atoms that made up the material effectively 'swallow' the energy of the incoming photon of light, causing those electrons to gain energy. But electrons can gain or lose energy only in discrete steps, known as 'quanta.' But the size of the steps varies from material to material. Silicon dioxide happens to have very large steps, which means that the energy from a single photon of light is not sufficient to bump up the electrons to the higher level of energy. Instead, the light passes through the material. (Most ultraviolet light, however, does have enough energy to be absorbed, which is why you can't get a suntan through a glass window.) But light doesn't simply pass through glass; it can also be bent and distorted or even broken up into its component wavelengths. Glass could be used to change the look of the world, by bending light in precise ways. This turned out to be even more revolutionary than simple transparency.

In the monasteries of the twelfth and thirteenth centuries, monks laboring over religious manuscripts in candlelit rooms used curved chunks of glass as a reading aid. They would run what were effectively bulky magnifiers over the page, enlarging the Latin inscriptions. No one is sure exactly when or where it happened, but somewhere around this time in Northern Italy, glassmakers came up with an innovation that would change the way we see the world, or at least clarify it: shaping glass into small disks that bulge in the center, placing each one in a frame, and joining the frames together at the top, creating the world's first spectacles.

Those early spectacles were called *roidi da ogli*, meaning 'disks for the eyes.' Thanks to their resemblance to lentil beans – *lentes* in Latin – the disks themselves came to be called 'lenses.' For several generations, these ingenious new devices were almost exclusively the province of monastic scholars. The condition of 'hyperopia' – farsightedness – was widely distributed through the population, but most people didn't notice that they suffered from it, because they didn't read. For a monk, straining to translate Lucretius by the flickering light of a candle, the need for spectacles was all too apparent. But the general population – the vast majority of them illiterate – had almost no occasion to discern tiny shapes like letterforms as part of their daily routine. People were farsighted; they just didn't have any real reason to notice that they were farsighted. And so spectacles remained rare and expensive objects.

What changed all of that, of course, was Gutenberg's invention of the printing press in the 1440s. You could fill a small library with the amount of historical scholarship that has been published documenting the impact of the printing press, the creation of what Marshall McLuhan famously called 'the Gutenberg galaxy.' Literacy rates rose dramatically; subversive scientific and religious theories routed around the official channels of orthodox belief; popular amusements like the novel and printed pornography became commonplace. But Gutenberg's great breakthrough had another, less celebrated effect: it made a massive number of people aware for the first time that they were farsighted. And that revelation created a surge in demand for spectacles.

What followed was one of the most extraordinary cases of the hummingbird effect in modern history. Gutenberg made printed books relatively cheap and portable, which triggered a rise in literacy, which exposed a flaw in the visual acuity of a sizable part of the population, which then created a new market for the manufacture of spectacles. Within a hundred years of Gutenberg's invention, thousands of spectacle makers around Europe were thriving, and glasses became the first piece of advanced technology – since the invention of clothing in Neolithic times – that ordinary people would regularly wear on their bodies.

But the coevolutionary dance did not stop there. Just as the nectar of flowering plants encouraged a new kind of flight in the hummingbird, the economic incentive created by the surging market for spectacles engendered a new pool of expertise. Europe was not just awash in lenses, but also in *ideas* about lenses. Thanks to the printing press, the Continent was suddenly populated by people who were experts at manipulating light through slightly convex pieces of glass. These were the hackers of the first optical revolution. Their experiments would inaugurate a whole new chapter in the history of vision.

In 1590 in the small town of Middleburg in the Netherlands, father and son spectacle makers Hans and Zacharias Janssen experimented with lining up two lenses, not side by side like spectacles, but in line with each other, magnifying the objects they observed, thereby inventing the microscope. Within seventy years, the British scientist Robert Hooke had published his groundbreaking illustrated volume *Micrographia*, with gorgeous hand-drawn images re-creating what Hooke had seen through his microscope. Hooke analyzed fleas, wood, leaves, even his own frozen urine. But his most influential discovery came by carving off a thin sheaf of cork and viewing it through the microscope lens. 'I could exceeding plainly perceive it to be all perforated and porous, much like a Honey-comb,' Hooke wrote, 'but that the pores of it were not regular; yet it was not unlike a Honey-comb in these particulars . . . these pores, or cells, were not very deep, but consisted of a great many little Boxes.' With that sentence, Hooke gave a name to one of life's fundamental building blocks – the cell – leading the way to a revolution in science and medicine. Before long the microscope would reveal the invisible colonies of bacteria and viruses that both sustain and threaten human life, which in turn led to modern vaccines and antibiotics.

The microscope took nearly three generations to produce truly transformative science, but for some reason the telescope generated its revolutions more quickly. Twenty years after the invention of the microscope, a cluster of Dutch lensmakers, including Zacharias Janssen, more or less simultaneously invented the telescope. (Legend has it that one of them, Hans Lippershey, stumbled upon the idea while watching his children playing with his lenses.) Lippershey was the first to apply for a patent, describing a device 'for seeing things far away as if they were nearby.' Within a year, Galileo got word of this miraculous new device, and modified the Lippershey design to reach a magnification of ten times normal vision. In January of 1610, just two years after Lippershey had filed for his patent, Galileo used the telescope to observe that moons were orbiting Jupiter, the first real challenge to the Aristotelian paradigm that assumed all heavenly bodies circled the Earth.

This is the strange parallel history of Gutenberg's invention. It has long been associated with the scientific revolution, for several reasons. Pamphlets and treatises from alleged heretics like Galileo could circulate ideas outside the censorious limits of the Church, ultimately undermining its authority; at the same time, the system of citation and reference that evolved in the decades after Gutenberg's Bible became an essential tool in applying the scientific method. But Gutenberg's creation advanced the march of science in another, less familiar way: it expanded possibilities of lens design, of glass itself. For the first time, the peculiar physical properties of silicon dioxide were not just being harnessed to let us see things that we could already see with our own eyes; we could now see things that transcended the natural limits of human vision.

The lens would go on to play a pivotal role in nineteenth- and twentieth-century media. It was first utilized by photographers to focus beams of light on specially treated paper that captured images, then by filmmakers to both record and subsequently project moving images for the first time. Starting in the 1940s, we began coating glass with phosphor and firing electrons at it, creating the hypnotic images of television. Within a few years, sociologists and media theorists were declaring that we had become a 'society of the image,' the literate Gutenberg galaxy giving way to the blue glow of the TV screen and the Hollywood glamour shot. Those transformations emerged out of a wide range of innovations and materials, but all of them, in one way or another, depended on the unique ability of glass to transmit and manipulate light.

To be sure, the story of the modern lens and its impact on media is not terribly surprising. There's an intuitive line that you can follow from the lenses of the first spectacles, to the lens of a microscope, to the lens of a camera. Yet glass would turn out to have another bizarre physical property, one that even the master glassblowers of Murano had failed to exploit.

As professors go, the physicist Charles Vernon Boys was apparently a lousy one. H. G. Wells, who was briefly one of Boys's students at London's Royal College of Science, later described him as 'one of the worst teachers who has ever turned his back on a restive audience . . . [He] messed about with the blackboard, galloped through an hour of talk, and bolted back to the apparatus in his private room.'

But what Boys lacked in teaching ability he made up for in his gift for experimental physics, designing and building scientific instruments. In 1887, as part of his physics experiments, Boys wanted to create a very fine shard of glass to measure the effects of delicate physical forces on objects. He had an idea that he could use a thin fiber of glass as a balance arm. But first he had to make one.

Hummingbird effects sometimes happen when an innovation in one field exposes a flaw in some other technology (or in the case of the printed book, in our own anatomy) that can be corrected only by another discipline altogether. But sometimes the effect arrives thanks to a different kind of breakthrough: a dramatic increase in our ability to *measure* something, and an improvement in the tools we build for measuring. New ways of measuring almost always imply new ways of making. Such was the case with Boys's balance arm. But what made Boys such an unusual figure in the annals of innovation is the decidedly unorthodox tool he used in pursuit of this new measuring device. To create his thin string of glass, Boys built a special crossbow in his laboratory, and created lightweight arrows (or bolts) for it. To one bolt he attached the end of a glass rod with sealing wax. Then he heated glass until it softened, and he fired the bolt. As the bolt hurtled toward its target, it pulled a tail of fiber from the molten glass clinging to the crossbow. In one of his shots, Boys produced a thread of glass that stretched almost ninety feet long.

'If I had been promised by a good fairy for anything I desired, I would have asked for one thing with so many valuable properties as these fibres,' Boys would later write. Most astonishing, though, was how strong the fiber was: as durable, if not more so, than an equivalently sized strand of steel. For thousands of years, humans had employed glass for its beauty and transparency, and tolerated its chronic fragility. But Boys's crossbow experiment suggested that there was one more twist in the story of this amazingly versatile material: using glass for its *strength*.

By the middle of the next century, glass fibers, now wound together in a miraculous new material called fiberglass, were everywhere: in home insulation, clothes, surfboards, megayachts, helmets, and the circuit boards that connected the chips of a modern computer. The fuselage of Airbus's flagship jet, the A380 – the largest commercial aircraft in the skies – is built with a composite of aluminum and fiberglass, making it much more resistant to fatigue and damage than traditional aluminum shells. Ironically, most of these applications ignored silicon dioxide's strange capacity to transmit light waves: most objects made of fiberglass do not look to the untutored eye to be made of glass at all. During the first decades of innovation with glass fibers, this emphasis on nontransparency made sense. It was useful to allow light to pass through a windowpane or a lens, but why would you need to pass light through a fiber not much bigger than a human hair?

The transparency of glass fibers became an asset only once we began thinking of light as a way to encode digital information. In 1970, researchers at Corning Glassworks – the Murano of modern times – developed a type of glass that was so extraordinarily clear that if you created a block of it the length of a bus, it would be just as transparent as looking through a normal windowpane. (Today, after further refinements, the block could be a half-mile long with the same clarity.) Scientists at Bell Labs then took fibers of this super-clear glass and shot laser beams down the length of them, fluctuating optical signals that corresponded to the zeroes and ones of binary code. This hybrid of two seemingly unrelated inventions – the concentrated, orderly light of lasers, and the hyper-clear glass fibers – came to be known as fiber optics. Using fiber-optic cables was vastly more efficient than sending electrical signals over copper cables, particularly for long distances: light allows much more bandwidth and is far less susceptible to noise and interference than is electrical energy. Today, the backbone of the global Internet is built out of fiber-optic cables. Roughly ten distinct cables traverse the Atlantic Ocean, carrying almost all the voice and data communications between the continents. Each of those cables contains a collection of separate fibers, surrounded by layers of steel and insulation to keep them watertight and protected from fishing trawlers, anchors, and even sharks. Each individual fiber is thinner than a piece of straw. It seems impossible, but the fact is that you can hold the entire collection of all the voice and data traffic traveling between North America and Europe in the palm of one hand. A thousand innovations came together to make that miracle possible: we had to invent the idea of digital data itself, and laser beams, and computers at both ends that could transmit and receive those beams of information – not to mention the ships that lay and repair the cables. But those strange bonds of silicon dioxide, once again, turn out to be central to the story. The World Wide Web is woven together out of threads of glass.

Think of that iconic, early-twenty-first-century act: snapping a selfie on your phone as you stand in some exotic spot on vacation, and then uploading the image to Instagram or Twitter, where it circulates to other people's phones and computers all around the world. We're accustomed to celebrating the innovations that have made this act almost second nature to us now: the miniaturization of digital computers into handheld devices, the creation of the Internet and the

Web, the interfaces of social-networking software. What we rarely do is recognize the way glass supports this entire network: we take pictures through glass lenses, store and manipulate them on circuit boards made of fiberglass, transmit them around the world via glass cables, and enjoy them on screens made of glass. It's silicon dioxide all the way down the chain.

It's easy to make fun of our penchant for taking selfies, but in fact there is a long and storied tradition behind that form of self-expression. Some of the most revered works of art from the Renaissance and early modernism are self-portraits; from Dürer to Leonardo, to Rembrandt, all the way to van Gogh with his bandaged ear, painters have been obsessed with capturing detailed and varied images of themselves on the canvas. Rembrandt, for instance, painted around forty self-portraits over the course of his life. But the interesting thing about self-portraiture is that it effectively doesn't exist as an artistic convention in Europe before 1400. People painted landscapes and royalty and religious scenes and a thousand other subjects. But they didn't paint themselves.

The explosion of interest in self-portraiture was the direct result of yet another technological breakthrough in our ability to manipulate glass. Back in Murano, the glassmakers had figured out a way to combine their crystal-clear glass with a new innovation in metallurgy, coating the back of the glass with an amalgam of tin and mercury to create a shiny and highly reflective surface. For the first time, mirrors became part of the fabric of everyday life. This was a revelation on the most intimate of levels: before mirrors came along, the average person went through life without ever seeing a truly accurate representation of his or her face, just fragmentary, distorted glances in pools of water or polished metals.

Mirrors appeared so magical that they were quickly integrated into somewhat bizarre sacred rituals: During holy pilgrimages, it became common practice for well-off pilgrims to take a mirror with them. When visiting sacred relics, they would position themselves so that they could catch sight of the bones in the mirror's reflection. Back home, they would then show off these mirrors to friends and relatives, boasting that they had brought back physical evidence of the relic by capturing the reflection of the sacred scene. Before turning to the printing press, Gutenberg had the start-up idea of manufacturing and selling small mirrors for departing pilgrims.

But the mirror's most significant impact would be secular, not sacred. Filippo Brunelleschi employed a mirror to invent linear perspective in painting, by drawing a reflection of the Florence Baptistry instead of his direct perception of it. The art of the late Renaissance is heavily populated by mirrors lurking inside paintings, most famously in Diego Velázquez's inverted masterpiece, *Las Meninas*, which shows the artist (and the extended royal family) in the middle of painting King Philip IV and Queen Mariana of Spain. The entire image is captured from the point of view of two royal subjects sitting for their portrait; it is, in a very literal sense, a painting about the act of painting. The king and queen are visible only in one small fragment of the canvas, just to the right of Velázquez himself: two small, blurry images reflected back in a mirror.

As a tool, the mirror became an invaluable asset to painters who could now capture the world around them in a far more realistic fashion, including the detailed features of their own faces. Leonardo da Vinci observed the following in his notebooks (using mirrors, naturally, to write in his legendary backward script):

> When you wish to see whether the general effect of your picture corresponds with that of the object represented after nature, take a mirror and set it so that it reflects the actual thing, and then compare the reflection with your picture, and consider carefully whether the subject of the two images is in conformity with both, studying especially the mirror. The mirror ought to be taken as a guide.

The historian Alan MacFarlane writes of the role of glass in shaping artistic vision, 'It is as if all humans had some kind of systematic myopia, but one which made it impossible to see, and particularly to represent, the natural world with precision and clarity. Humans normally saw nature symbolically, as a set of signs . . . What glass ironically did was to take away or compensate for the dark glass of human sight and the distortions of the mind, and hence to let in more light.'

At the exact moment that the glass lens was allowing us to extend our vision to the stars or microscopic cells, glass mirrors were allowing us to see ourselves for the first time. It set in motion a reorientation of society that was more subtle, but no less transformative, than the reorientation of our place in the universe that the telescope engendered. 'The most powerful prince in the world created a vast hall of mirrors, and the mirror spread from one room to another in the bourgeois household,' Lewis Mumford writes in his *Technics and Civilization*. 'Self-consciousness, introspection, mirror-conversation developed with the new object itself.' Social conventions as well as property rights and other legal customs began to revolve around the individual rather than the older, more collective units: the family, the tribe, the city, the kingdom. People began writing about their interior lives with far more scrutiny. Hamlet ruminated onstage; the novel emerged as a dominant form of storytelling, probing the inner mental lives of its characters with an unrivaled depth. Entering a novel, particularly a first-person narrative, was a kind of conceptual parlor trick: it let you swim through the consciousness, the thoughts and emotions, of other people more effectively than any aesthetic form yet invented. The psychological novel, in a sense, is the kind of story you start wanting to hear once you begin spending meaningful hours of your life staring at yourself in the mirror.

How much does this transformation owe to glass? Two things are undeniable: the mirror played a direct role in allowing artists to paint themselves and invent perspective as a formal device; and shortly thereafter a fundamental shift occurred in the consciousness of Europeans that oriented them around the self in a new way, a shift that would ripple across the world (and that is still rippling). No doubt many forces converged to make this shift possible: the self-centered world played well with the early forms of modern capitalism that were thriving in places like Venice and Holland (home to those masters of painterly introspection, Dürer and Rembrandt). Likely, these various forces complemented each other: glass mirrors were among the first high-tech furnishings for the home, and once we began gazing into those mirrors, we began to see ourselves differently, in ways that encouraged the market systems that would then happily sell us more mirrors. It's not that the mirror made the Renaissance, exactly, but that it got caught up in a positive feedback loop with other social forces, and its unusual capacity to reflect light strengthened those forces. This is what the robot historian's perspective allows us to see: the technology is not a single cause of a cultural transformation like the Renaissance, but it is, in many ways, just as important to the story as the human visionaries that we conventionally celebrate.

McFarlane has an artful way of describing this kind of causal relationship. The mirror doesn't 'force' the Renaissance to happen; it 'allows' it to happen. The elaborate reproductive strategy of the pollinators didn't *force* the hummingbird to evolve its spectacular aerodynamics; it created the conditions that *allowed* the hummingbird to take advantage of flower's free sugars by evolving such a distinctive trait. The fact that the hummingbird is so unique in the avian kingdom suggests that, had the flowers not evolved their symbiotic dance with the insects, the hummingbird's hovering skills would have never come into being. It's easy to imagine a world with flowers but without hummingbirds. But it's much harder to imagine a world without flowers but *with* hummingbirds.

The same holds true for technological advances like the mirror. Without a technology that enabled humans to see a clear reflection of reality, including their own faces, the particular constellation of ideas in art and philosophy and politics that we call the Renaissance would have had a much more difficult time coming into being. (Japanese culture had highly prized steel mirrors during roughly the same period, but never adopted them for the same introspective use that flourished in Europe – perhaps in part because steel reflected much less light than glass mirrors, and added unnatural coloring to the image.) Yet the mirror was not exclusively dictating the terms of the European revolution in the sense of self. A different culture, inventing the fine glass mirror at a different point in its historical development, might not have experienced the same intellectual revolution, because the rest of its social order differed from that of fifteenth-century Italian hill-towns. The Renaissance also benefited from a patronage system that enabled its artists and scientists to spend their days playing with mirrors instead of, say, foraging for nuts and berries. A Renaissance without the Medici – not the individual family, of course, but the economic class they represent – is as hard to imagine as the Renaissance without the mirror.

It should probably be said that the virtues of the society of the self are entirely debatable. Orienting laws around individuals led directly to an entire tradition of human rights and the prominence of individual liberty in legal codes. That has to count as progress. But reasonable people disagree about whether we have now tipped the scales too far in the direction of individualism, away from those collective organizations: the union, the community, the state. Resolving those disagreements requires a different set of arguments – and values – than the ones we need to explain where those disagreements came from. The mirror helped invent the modern self, in some real but unquantifiable way.

That much we should agree on. Whether that was a good thing in the end is a separate question, one that may never be settled conclusively.

The dormant volcano of Mauna Kea on Hawaii's Big Island rises almost fourteen thousand feet above sea level, though the mountain extends almost another twenty thousand feet down to the ocean floor below, making it significantly larger than Mount Everest in terms of base-to-peak height. It is one of the few places in the world where you can drive from sea level to fourteen thousand feet in a matter of hours. At the summit, the landscape is barren, almost Martian, in its rocky, lifeless expanse. An inversion layer generally keeps clouds several thousand feet below the volcano's peak; the air is as dry as it is thin. Standing on the summit, you are as far from the continents of earth as you can be while standing on land, which means the atmosphere around Hawaii – undisturbed by the turbulence of the sun's energy bouncing off or being absorbed by large, varied landmasses – is as stable as just about anywhere on the planet. All of these properties make the peak of Mauna Kea one of the most otherworldly places you can visit. Appropriately enough, they also make it a sublime location for stargazing.

Today, the summit of Mauna Kea is crowned by thirteen distinct observatories, massive white domes scattered across the red rocks like some gleaming outpost on a distant planet. Included in this group are the twin telescopes of the W. M. Keck Observatory, the most powerful optical telescopes on earth. The Keck telescopes would seem to be a direct descendant of Hans Lippershey's creation, only they do not rely on lenses to do their magic. To capture light from distant corners of the universe, you would need lenses the size of a pickup truck; at that size, glass becomes difficult to physically support and introduces inevitable distortions into the image. And so, the scientists and engineers behind Keck employed another technique to capture extremely faint traces of light: the mirror.

Each telescope has thirty-six hexagonal mirrors that together become a twenty-foot-wide reflective canvas. That light is reflected up to a second mirror and then down to a set of instruments, where the images can then be processed and visualized on a computer screen. (There is no vantage point at Keck where one can gaze directly through the telescope the way Galileo and countless astronomers since have done.) But even in the thin, ultra-stable atmosphere above Mauna Kea, small disturbances can blur the images captured by Keck. And so the observatories employ an ingenious system called 'adaptive optics' to correct the vision of the telescopes. Lasers are beamed into the night sky above Keck, effectively creating an artificial star in the heavens. That false star becomes a kind of reference point; because the scientists know exactly what the laser should look like in the heavens were there no atmospheric distortion, they are able to get a measurement of the existing distortion by comparing the 'ideal' laser image and what the telescopes actually register. Guided by that map of atmospheric noise, computers instruct the mirrors of the telescope to flex slightly based on the exact distortions in the skies above Mauna Kea that night. The effect is almost exactly like putting spectacles on a nearsighted person: distant objects suddenly become significantly clearer.

Of course, with the Keck telescopes, those distant objects are galaxies and supernovas that are, in some cases, billions of light-years away. When we look through the mirrors of Keck, we are looking into the distant past. Once again, glass has extended our vision: not just down to the invisible world of cells and microbes, or the global connectivity of the cameraphone, but all the way back to the early days of the universe. Glass started out as trinkets and empty vessels. A few thousand years later, perched above the clouds at the top of Mauna Kea, it has become a time machine.

The story of glass reminds us how our ingenuity is both confined and empowered by the physical properties of the elements around us. When we think of the entities that made the modern world, we usually talk about the great visionaries of science and politics, or breakthrough inventions, or large collective movements. But there is a material element to our history as well: not the dialectical materialism that Marxism practiced, where 'material' meant the class struggle and the ultimate primacy of economic explanations. Material history, instead, in the sense of history as shaped by the basic building blocks of matter, which are then connected to things like social movements or economic systems. Imagine you could rewrite the Big Bang (or play God, depending on your metaphor) and create a universe that was exactly like ours, with only one tiny change: those electrons on the silicon atom don't behave quite the same way. In this alternate universe, the electrons *absorb* light like most materials, instead of letting the photons pass through them. Such a small adjustment might well have made no difference at all for the entire evolution of *Homo sapiens* until a few thousand years ago. But then, amazingly, everything changed. Humans began exploiting the quantum behavior of those silicon electrons in countless different ways. On some fundamental level, it is impossible to imagine the last millennium without transparent glass. We can now manipulate carbon (in the form of that defining twentieth-century compound, plastic) into durable transparent materials that can do the job of glass, but that expertise is less than a century old. Tweak those silicon electrons, and you rob the last thousand years of windows, spectacles, lenses, test tubes, lightbulbs. (High-quality mirrors might have been independently invented using other reflective materials, though it would likely have taken a few centuries longer.) A world without glass would not just transform the edifices of civilization, by removing all the stained-glass windows of the great cathedrals and the sleek, reflective surfaces of the modern cityscape. A world without glass would strike at the foundation of modern progress: the extended life spans that come from understanding the cell, the virus, and the bacterium; the genetic knowledge of what makes us human; the astronomer's knowledge of our place in the universe. No material on Earth mattered more to those conceptual breakthroughs than glass.

In a letter to a friend about the book of natural history that he never got around to writing, René Descartes described how he had wanted to tell the story of glass: 'How from these ashes, by the mere intensity of [heat's] action, it formed glass: for as this transmutation of ashes into glass appeared to me as wonderful as any other nature, I took a special pleasure in describing it.' Descartes was close enough to the original glass revolution to perceive its magnitude. Today, we are too many steps away from the material's original influence to appreciate just how important it was, and continues to be, to everyday existence.

This is one of those places where the long-zoom approach illuminates, allowing us to see things that we would have otherwise missed had we focused on the usual suspects of historical storytelling. Invoking the physical elements in discussing historical change is not unheard of, of course. Most of us accept the idea that carbon has played an essential role in human activity since the industrial revolution. But in a way, this is not really news: carbon has been essential to just about everything living organisms have done since the primordial soup. But humans didn't have much use for silicon dioxide until the glassmakers began to tinker with its curious properties a thousand years ago. Today, if you look around the room you're currently occupying, there might easily be a hundred objects within reach that depend on silicon dioxide for their existence, and even more that rely on the element silicon itself: the panes of glass in your windows or skylights, the lens in your cameraphone, the screen of your computer, everything with a microchip or a digital clock. If you were casting starring roles for the chemistry of daily life ten thousand years ago, the top billing would be the same as it is today: we're heavy users of carbon, hydrogen, oxygen. But silicon wouldn't likely have even received a credit. While silicon is abundant on Earth – more than 90 percent of the crust is made up of the element – it plays almost no role in the natural metabolisms of life-forms on the planet. Our bodies are dependent on carbon, and many of our technologies (fossil fuels and plastics) display the same dependence. But the need for silicon is a modern craving.

The question is: Why did it take so long? Why were the extraordinary properties of this substance effectively ignored by nature, and why did those properties suddenly become essential to human society starting roughly a thousand years ago? In trying to address these questions, of course, we can only speculate. But surely one answer has to do with another technology: the furnace. One reason that evolution didn't find much use for silicon dioxide is that most of the really interesting things about the substance don't appear until you get over 1,000 degrees Fahrenheit. Liquid water and carbon do wonderfully inventive things at the earth's atmospheric temperature, but it's hard to see the promise of silicon dioxide until you can melt it, and the earth's environment – at least on the surface of the planet – simply doesn't get that hot. This was the hummingbird effect that the furnace unleashed: by learning how to generate extreme heat in a controlled environment, we unlocked the molecular potential of silicon dioxide, which soon transformed the way we see the world, and ourselves.

In a strange way, glass was trying to extend our vision of the universe from the very beginning, way before we were smart enough to notice. Those glass fragments from the Libyan Desert that made it into King Tut's tomb had puzzled archeologists, geologists, and astrophysicists alike for decades. The semiliquid molecules of silicon dioxide suggested that they had formed at temperatures that could only have been created by a direct meteor strike, and yet there was no evidence of an impact crater anywhere in the vicinity. So where had those extraordinary temperatures come from? Lightning can strike a small patch of silica with glassmaking heat, but it can't strike acres of sand in a single blast. And so scientists began to explore the idea that the Libyan glass arose from a comet colliding with the earth's atmosphere and exploding over the desert sands. In 2013, a South African geochemist named Jan Kramers analyzed a mysterious pebble from the site and determined that it had originated in the nucleus of a comet, the first such object to be discovered on Earth. Scientists and space agencies have spent billions of dollars searching for particles of comets because they offer such profound insight into the formation of solar systems. The pebble from the Libyan Desert now gives them direct access to the geochemistry of comets. And all the while, glass was pointing the way.

CHAPTER 2

Cold

In the early summer months of 1834, a three-masted bark vessel named the *Madagascar* sailed into the port of Rio de Janeiro, its hull filled with the most implausible of cargo: a frozen New England lake. The *Madagascar* and her crew were in the service of an enterprising and dogged Boston businessman named Frederic Tudor. History now knows him as 'the Ice King,' but for most of his early adulthood he was an abject failure, albeit one with remarkable tenacity.

'Ice is an interesting subject for contemplation,' Thoreau wrote in *Walden*, gazing out at the 'beautifully blue' frozen expanse of his Massachusetts pond. Tudor had grown up contemplating the same scenery. As a well-to-do young Bostonian, his family had long enjoyed the frozen water from the pond on their country estate, Rockwood – not just for its aesthetics, but also for its enduring capacity to keep things cold. Like many wealthy families in northern climes, the Tudors stored blocks of frozen lake water in icehouses, two-hundred-pound ice cubes that would remain marvelously unmelted until the hot summer months arrived, and a new ritual began: chipping off slices from the blocks to freshen drinks, make ice cream, cool down a bath during a heat wave.

The idea of a block of ice surviving intact for months without the benefit of artificial refrigeration sounds unlikely to the modern ear. We are used to ice preserved indefinitely thanks to the many deep-freeze technologies of today's world. But ice in the wild is another matter – other than the occasional glacier, we assume that a block of ice can't survive longer than an hour in summer heat, much less months.

But Tudor knew from personal experience that a large block of ice could last well into the depths of summer if it was kept out of the sun – or at least it would last through the late spring of New England. And that knowledge would plant the seed of an idea in his mind, an idea that would ultimately cost him his sanity, his fortune, and his freedom – before it made him an immensely wealthy man.

At the age of seventeen, Tudor's father sent him on a voyage to the Caribbean, accompanying his older brother John, who suffered from a knee ailment that had effectively rendered him an invalid. The idea was that the warm climates would improve John's health, but in fact they had the opposite effect: arriving in Havana, the Tudor brothers were quickly overwhelmed by the muggy weather. They soon sailed north back to the mainland, stopping in Savannah and Charleston, but the early summer heat followed them, and John fell ill with what may have been tuberculosis. Six months later, he was dead at the age of twenty.

As a medical intervention, the Tudor brothers' Caribbean adventure was a complete disaster. But suffering through the inescapable humidity of the tropics in the full regalia of a nineteenth-century gentleman suggested a radical – some would say preposterous – idea to young Frederic Tudor: if he could somehow transport ice from the frozen north to the West Indies, there would be an immense market for it. The history of global trade had clearly demonstrated that vast fortunes could be made by transporting a commodity that was ubiquitous in one environment to a place where it was scarce. To the young Tudor, ice seemed to fit the equation perfectly: nearly worthless in Boston, ice would be priceless in Havana.

The ice trade was nothing more than a hunch, but for some reason Tudor kept it alive in his mind, through the grieving after his brother's demise, through the aimless years of a young man of means in Boston society. Sometime during this period, two years after his brother's death, he shared his implausible scheme with his brother William, and his future brother-in-law, the even wealthier Robert Gardiner. A few months after his sister's wedding, Tudor began taking notes in a journal. As a frontispiece, he drew a sketch of the Rockwood building that had long enabled his family to escape the warmth of the summer sun. He called it the 'Ice House Diary.' The first entry read: 'Plan etc for transporting Ice to Tropical Climates. Boston Augst 1st 1805 William and myself have this day determined to get together what property we have and embark in the undertaking of carrying ice to the West Indies the ensuing winter.'

The entry was typical of Tudor's whole demeanor: brisk, confident, almost comically ambitious. (Brother William was apparently less convinced of the scheme's promise.) Tudor's confidence in his scheme derived from the ultimate value of the ice once it made its way to the tropics: 'In a country where at some seasons of the year the heat is almost unsupportable,' he wrote in a subsequent entry, 'where at times the common necessary of life, water, cannot be had but in a tepid state – Ice must be considered as out doing most other luxuries.' The ice trade was destined to endow the Tudor brothers with 'fortunes larger than we shall know what to do with.' He seems to have given less thought to the challenges of transporting the ice. In correspondence from the period, Tudor relays thirdhand stories – almost certainly apocryphal – of ice cream being shipped intact from England to Trinidad as prima facie evidence that his plan would work. Reading the 'Ice House Diary' now, you can hear the voice of a young man in the full fever of conviction, closing the cognitive blinds against doubt and counterargument.

However deluded Frederic might have seemed, he had one thing in his favor: he had the means to put the broad strokes of his plan in motion. He had enough money to hire a ship, and an endless supply of ice, manufactured by Mother Nature each winter. And so, in November 1805, Tudor dispatched his brother and cousin off to Martinique as an advance guard, with instructions to negotiate exclusive rights to the ice that would follow several months later. While waiting for word from his envoys, Tudor bought a brig called the *Favorite* for $4,750 and began harvesting ice in preparation for the journey. In February, Tudor set sail from Boston Harbor, the *Favorite* loaded with a full cargo of Rockwood ice, bound for the West Indies. Tudor's scheme was bold enough to attract the attentions of the press, though the tone left something to be desired. 'No joke,' the *Boston Gazette* reported. 'A vessel with a cargo of 80 tons of Ice has cleared out from this port for Martinique. We hope this will not prove to be a slippery speculation.'

The *Gazette*'s derision would turn out to be well founded, though not for the reasons one might expect. Despite a number of weather-related delays, the ice survived the journey in remarkably good shape. The problem proved to be one that Tudor had never contemplated. The residents of Martinique had no interest in his exotic frozen bounty. They simply had no idea what to do with it.

We take it for granted in the modern world that an ordinary day will involve exposure to a wide range of temperatures. We enjoy piping hot coffee in the morning and ice cream for dessert at the end of the day. Those of us who live in climates with hot summers expect to bounce back and forth between air-conditioned offices and brutal humidity; where winter rules, we bundle up and venture out into the frigid streets, and turn up the thermostat when we return home. But the overwhelming majority of humans living in equatorial climes in 1800 would have literally never once experienced anything cold. The idea of frozen water would have been as fanciful to the residents of Martinique as an iPhone.

The mysterious, almost magical, properties of ice would eventually appear in one of the great opening lines of twentieth-century literature, in Gabriel García Márquez's *One Hundred Years of Solitude*: 'Many years later, as he faced the firing squad, Colonel Aureliano Buendía was to remember that distant afternoon when his father took him to discover ice.' Buendía recalls a series of fairs put on by roving gypsies during his childhood, each showcasing some extraordinary new technology. The gypsies display magnetic ingots, telescopes, and microscopes; but none of these engineering achievements impress the residents of the imaginary South American town of Macondo as much as a simple block of ice.

But sometimes the sheer novelty of an object can make its utility hard to discern. This was Tudor's first mistake. He assumed the absolute novelty of ice would be a point in his favor. He figured his blocks of ice would 'out-do' all the other luxuries. Instead, they just received blank stares.

The indifference to ice's magical powers had prevented Tudor's brother William from lining up an exclusive buyer for the cargo. Even worse, William had failed to establish a suitable location to store the ice. Tudor had made it all the way to Martinique but found himself with no demand for a product that was melting in the tropical heat at an alarming rate. He posted handbills around town that included specific instructions on how to carry and preserve the ice, but found few takers. He did manage to make some ice cream, thereby impressing a few locals who believed the delicacy couldn't be created so close to the equator. But the trip was ultimately a complete failure. In his diary, he estimated that he had lost nearly $4,000 with his tropical misadventure.

The bleak pattern of the Martinique voyage would repeat itself in the years to come, with ever more catastrophic results. Tudor sent a series of ice ships to the Caribbean, with only a modest increase in demand for his product. In the meantime, his family fortunes collapsed, and the Tudors retreated to their Rockwood farm, which like most New

England land had very poor agricultural prospects. Harvesting the ice was the family's last best hope. But it was a hope that most of Boston openly ridiculed, and a series of shipwrecks and embargoes made that ridicule seem increasingly appropriate. In 1813, Tudor was thrown in debtor's prison. He penned the following entry in his diary several days later:

> On Monday the 9th instant I was arrested . . . and locked up as a debtor in Boston jail . . . On this memorable day in my little annals I am 28 years 6 months and 5 days old. It is an event which I think I could not have avoided: but it is a climax which I did hope to have escaped as my affairs are looking well at last after a fearful struggle with adverse circumstances for seven years – but it has taken place and I have endeavoured to meet it as I would the tempest of heaven which should serve to strengthen rather than reduce the spirit of a true man.

Tudor's fledgling business suffered from two primary liabilities. He had a demand problem, in that most of his potential customers didn't understand why his product might be useful. And he had a storage problem: he was losing too much of his product to the heat, particularly once it arrived in the tropics. But his New England base gave him one crucial advantage, beyond the ice itself. Unlike the U.S. South, with its sugar plantations and cotton fields, the northeastern states were largely devoid of natural resources that could be sold elsewhere. This meant that ships tended to leave Boston harbor empty, heading off for the West Indies to fill their hulls with valuable cargo before returning to the wealthy markets of the eastern seaboard. Paying a crew to sail a ship with no cargo was effectively burning money. Any cargo was better than nothing, which meant that Tudor could negotiate cheaper rates for himself by loading his ice onto what would have otherwise been an empty ship, and thereby avoiding the need to buy and maintain his own vessels.

Part of the beauty of ice, of course, was that it was basically free: Tudor needed only to pay workers to carve blocks of it out of the frozen lakes. New England's economy generated another product that was equally worthless: sawdust – the primary waste product of lumber mills. After years of experimenting with different solutions, Tudor discovered that sawdust made a brilliant insulator for his ice. Blocks layered on top of each other with sawdust separating them would last almost twice as long as unprotected ice. This was Tudor's frugal genius: he took three things that the market had effectively priced at zero – ice, sawdust, and an empty vessel – and turned them into a flourishing business.

Tudor's initial catastrophic trip to Martinique had made it clear that he needed on-site storage in the tropics that he could control; it was too dangerous to keep his rapidly melting product in buildings that weren't specifically engineered to insulate ice from the summer heat. He tinkered with multiple icehouse designs, finally settling on a double-shelled structure that used the air between two stone walls to keep the interior cool.

Tudor didn't understand the molecular chemistry of it, but both the sawdust and the double-shelled architecture revolved around the same principle. For ice to melt, it needs to pull heat from the surrounding environment to break the tetrahedral bonding of hydrogen atoms that gives ice its crystalline structure. (The extraction of heat from the surrounding atmosphere is what grants ice its miraculous capacity to cool us down.) The only place that heat exchange can happen is at the surface of the ice, which is why large blocks of ice survive for so long – all the interior hydrogen bonds are perfectly insulated from the exterior temperature. If you try to protect ice from external warmth with some kind of substance that conducts heat efficiently – metal for instance – the hydrogen bonds will break down quickly into water. But if you create a buffer between the external heat and the ice that conducts heat poorly, the ice will preserve its crystalline state for much longer. As a thermal conductor, air is about two thousand times less efficient than metal, and more than twenty times less efficient than glass. In his icehouses, Tudor's double-shelled structure created a buffer of air that kept the summer heat away from the ice; his sawdust packaging on the ships ensured that there were countless pockets of air between the wood shavings to keep the ice insulated. Modern insulators such as Styrofoam rely on the same technique: the cooler you take on a picnic keeps your watermelon chilled because it is made of polystyrene chains interspersed with tiny pockets of gas.

By 1815, Tudor had finally assembled the key pieces of the ice puzzle: harvesting, insulation, transport, and storage. Still pursued by his creditors, he began making regular shipments to a state-of-the-art icehouse he had built in Havana, where an appetite for ice cream had been slowly maturing. Fifteen years after his original hunch, Tudor's ice trade had finally turned a profit. By the 1820s, he had icehouses packed with frozen New England water all over the

American South. By the 1830s, his ships were sailing to Rio and Bombay. (India would ultimately prove to be his most lucrative market.) By his death in 1864, Tudor had amassed a fortune worth more than $200 million in today's dollars.

Three decades after his first failed voyage, Tudor wrote these lines in his journal:

> This day I sailed from Boston thirty years ago in the Brig Favorite Capt Pearson for Martinique: with the first cargo of ice. Last year I shipped upwards of 30 cargoes of Ice and as much as 40 more were shipped by other persons . . . The business is established. It cannot be given up now and does not depend upon a single life. Mankind will have the blessing for ever whether I die soon or live long.

Tudor's triumphant (if long-delayed) success selling ice around the world seems implausible to us today not just because it's hard to imagine blocks of ice surviving the passage from Boston to Bombay. There's an additional, almost philosophical, curiosity to the ice business. Most of the trade in natural goods involves material that thrives in high-energy environments. Sugarcane, coffee, tea, cotton – all these staples of eighteenth- and nineteenth-century commerce were dependent on the blistering heat of tropical and subtropical climates; the fossil fuels that now circle the planet in tankers and pipelines are simply solar energy that was captured and stored by plants millions of years ago. You could make a fortune in 1800 by taking things that grew only in high-energy environments and shipping them off to low-energy climates. But the ice trade – arguably for the only time in the history of global commerce – reversed that pattern. What made ice valuable was precisely the low-energy state of a New England winter, and the peculiar capacity of ice to store that lack of energy for long periods of time. The cash crops of the tropics caused populations to swell in climates that could be unforgivingly hot, which in turn created a market for a product that allowed you to escape the heat. In the long history of human commerce, energy had always correlated with value: the more heat, the more solar energy, the more you could grow. But in a world that was tilting toward the productive heat of sugarcane and cotton plantations, cold could be an asset as well. That was Tudor's great insight.

In the winter of 1846, Henry Thoreau watched ice cutters employed by Frederic Tudor carve blocks out of Walden Pond with a horse-drawn plow. It might have been a scene out of Brueghel, men working in a wintry landscape with simple tools, far from the industrial age that thundered elsewhere. But Thoreau knew their labor was attached to a wider network. In his diaries, he wrote a lilting reverie on the global reach of the ice trade:

> Thus it appears that the sweltering inhabitants of Charleston and New Orleans, of Madras and Bombay and Calcutta, drink at my well . . . The pure Walden water is mingled with the sacred water of the Ganges. With favoring winds it is wafted past the site of the fabulous islands of Atlantis and the Hesperides, makes the periplus of Hanno, and, floating by Ternate and Tidore and the mouth of the Persian Gulf, melts in the tropic gales of the Indian seas, and is landed in ports of which Alexander only heard the names.

If anything, Thoreau was underestimating the scope of that global network – because the ice trade that Tudor created was about much more than frozen water. The blank stares that had confronted Tudor's first shipment of ice to Martinique slowly but steadily gave way to an ever widening dependence on ice. Ice-chilled drinks became a staple of life in southern states. (Even today, Americans are far more likely to enjoy ice with their beverages than Europeans, a distant legacy of Tudor's ambition.) By 1850, Tudor's success had inspired countless imitators, and more than a hundred thousand tons of Boston ice were shipped around the world in a single year. By 1860, two out of three New York homes had daily deliveries of ice. One contemporary account describes how tightly bound ice had become to the rituals of daily life:

> In workshops, composing rooms, counting houses, workmen, printers, clerks club to have their daily supply of ice. Every office, nook or cranny, illuminated by a human face, is also cooled by the presence of his crystal friend . . . It is as good as oil to the wheel. It sets the whole human machinery in pleasant action, turns the wheels of commerce, and propels the energetic business engine.

The dependence on natural ice became so severe that every decade or so an unusually warm winter would send the newspapers into a frenzy with speculation about an 'ice famine.' As late as 1906, the *New York Times* was running alarming headlines: 'Ice Up To 40 Cents And A Famine In Sight.' The paper went on to provide some historical context: 'Not in sixteen years has New York faced such an iceless prospect as this year. In 1890 there was a great deal of trouble and the whole country had to be scoured for ice. Since then, however, the needs for ice have grown vastly, and a famine is a much more serious matter now than it was then.' In less than a century, ice had gone from a curiosity to a luxury to a necessity.

Ice-powered refrigeration changed the map of America, nowhere more so than in the transformation of Chicago. Chicago's initial burst of growth had come after the nexus of canals and rail lines connected the city to both the Gulf of Mexico and the cities of the eastern seaboard. Its fortuitous location as a transportation hub – created both by nature and some of the most ambitious engineering of the century – enabled wheat to flow from the bountiful plains to the Northeast population centers. But meat couldn't make the journey without spoiling. Chicago developed a large trade in preserved pork starting in the middle of the century, with the first stockyards slaughtering the hogs on the outskirts of the city before sending the goods east in barrels. But fresh beef remained largely a local delicacy.

But as the century progressed, a supply/demand imbalance developed between the hungry cities of the Northeast and the cattle of the Midwest. As immigration fueled the population of New York and Philadelphia and other urban centers in the 1840s and 1850s, the supply of local beef failed to keep up with the surging demand in the growing cities. Meanwhile, the conquest of the Great Plains had enabled ranchers to breed massive herds of cattle, without a corresponding population base of humans to feed. You could ship live cattle by train to the eastern states to be slaughtered locally, but transporting entire cows was expensive, and the animals were often malnourished or even injured en route. Almost half would be inedible by the time they arrived in New York or in Boston.

It was ice that ultimately provided a way around this impasse. In 1868, the pork magnate Benjamin Hutchinson built a new packing plant, featuring 'cooling rooms packed with natural ice that allowed them to pack pork year-round, one of the principal innovations in the industry,' according to Donald Miller, in his history of nineteenth-century Chicago, *City of the Century*. It was the beginning of a revolution that would transform not only Chicago but the entire natural landscape of middle America. In the years after the fire of 1871, Hutchinson's cooling rooms would inspire other entrepreneurs to integrate ice-cooled facilities to the meatpacking trade. A few began transporting beef back east in open-air railcars during winter, relying on the ambient temperature to keep the steaks cold. In 1878, Gustavus Franklin Swift hired an engineer to build an advanced refrigerator car, designed from the ground up to transport beef to the eastern seaboard year round. Ice was placed in bins above the meat; at stops along the route, workers could swap in new blocks of ice from above, without disturbing the meat below. 'It was this application of elementary physics,' Miller writes, 'that transformed the ancient trade of beef slaughtering from a local to an international business, for refrigerator cars led naturally to refrigerator ships, which carried Chicago beef to four continents.' The success of that global trade transformed the natural landscape of the American plains in ways that are still visible today: the vast, shimmering grasslands replaced by industrial feedlots, creating, in Miller's words, 'a city-country [food] system that was the most powerful environmental force in transforming the American landscape since the Ice Age glaciers began their final retreat.'

The Chicago stockyards that emerged in the last two decades of the nineteenth century were, as Upton Sinclair wrote, 'the greatest aggregation of labor and capital ever gathered in one place.' Fourteen million animals were slaughtered in an average year. In many ways, the industrial food complex held in such disdain by modern-day 'slow food' advocates begins with the Chicago stockyards and the web of ice-cooled transport that extended out from those grim feedlots and slaughterhouses. Progressives like Upton Sinclair painted Chicago as a kind of Dante's Inferno of industrialization, but in reality, most of the technology employed in the stockyards would have been recognizable to a medieval butcher. The most advanced form of technology in the whole chain was the refrigerated railcar. Theodore Dreiser got it right when he described the stockyard assembly line as 'a direct sloping path to death, dissection, and the refrigerator.'

The conventional story about Chicago is that it was made possible thanks to the invention of the railroad and the building of the Erie Canal. But those accounts tell only part of the story. The runaway growth of Chicago would have never been possible without the peculiar chemical properties of water: its capacity for storing and slowly releasing cold with only the slightest of human interventions. If the chemical properties of liquid water had somehow turned out to be different, life on earth would have taken a radically different shape (or more likely, would not have evolved at all). But if water hadn't also possessed its peculiar aptitude for freezing, the trajectory of nineteenth-century America would have almost certainly been different as well. You could send spices around the globe without the advantages of refrigeration, but you couldn't send beef. Ice made a new kind of food network imaginable. We think of Chicago as a city of broad shoulders, of railroad empires and slaughterhouses. But it is just as true to say that it was built on the tetrahedral bonds of hydrogen.

If you widen your frame of reference, and look at the ice trade in the context of technological history, there is something puzzling, almost anachronistic, about Tudor's innovation. This was the middle of the nineteenth century, after all, an era of coal-powered factories, with railroads and telegraph wires connecting massive cities. And yet the state of the art in cold technology was still entirely based on cutting chunks of frozen water out of a lake. Humans had been experimenting with the technology of heat for at least a hundred thousand years, since the mastery of fire – arguably *Homo sapiens'* first innovation. But the opposite end of the thermal spectrum was much more challenging. A century into the industrial revolution, artificial cold was still a fantasy.

But the commercial demand for ice – all those millions of dollars flowing upstream from the tropics to the ice barons of New England – sent a signal out across the world that there was money to be made from cold, which inevitably sent some inventive minds off in search of the next logical step of artificial cold. You might assume Tudor's success would inspire a new generation of equally mercenary entrepreneur-inventors to create the revolution in man-made refrigeration. Yet, however much we may celebrate the start-up culture of today's tech world, essential innovations don't always come out of private-sector exploration. New ideas are not always motivated, like Tudor's, by dreams of 'fortunes larger than we shall know what to do with.' The art of human invention has more than one muse. While the ice trade began with a young man's dream of untold riches, the story of artificial cold began with a more urgent and humanitarian need: a doctor trying to keep his patients alive.

It's a story that begins at the scale of insects: in Apalachicola, Florida, a town of ten thousand people living alongside a swamp in a subtropical climate – the perfect environment for breeding mosquitoes. In 1842, abundant mosquitoes meant, inevitably, the risk of malaria. At the modest local hospital, a doctor named John Gorrie sat helpless as dozens of his patients burned up with fever.

Desperate for a way to reduce his patients' fevers, Gorrie tried suspending blocks of ice from the hospital ceiling. It turned out to be an effective solution: the ice blocks cooled the air; the air cooled the patients. With fevers reduced, some of his patients survived their illnesses. But Gorrie's clever hack, designed to combat the dangerous effects of subtropical climates, was ultimately undermined by another by-product of the environment. The tropical humidity that made Florida such a hospitable climate for mosquitoes also helped breed another threat: hurricanes. A string of shipwrecks delayed ice shipments from Tudor's New England, which left Gorrie without his usual supply.

And so the young doctor began mulling over a more radical solution for his hospital: making his own ice. Luckily for Gorrie, it happened to be the perfect time to have this idea. For thousands of years, the idea of making artificial cold had been almost unthinkable to human civilization. We invented agriculture and cities and aqueducts and the printing press, but cold was outside the boundaries of possibility for all those years. And yet somehow artificial cold became imaginable in the middle of the nineteenth century. To use the wonderful phrase of the complexity theorist Stuart Kauffman, cold became part of the 'adjacent possible' of that period.

How do we explain this breakthrough? It's not just a matter of a solitary genius coming up with a brilliant invention because he or she is smarter than everyone else. And that's because ideas are fundamentally *networks* of other ideas. We take the tools and metaphors and concepts and scientific understanding of our time, and we remix them into something new. But if you don't have the right building blocks, you can't make the breakthrough, however brilliant you might be. The smartest mind in the world couldn't invent a refrigerator in the middle of the seventeenth century. It simply wasn't part of the adjacent possible at that moment. But by 1850, the pieces had come together.

The first thing that had to happen seems almost comical to us today: we had to discover that air was actually made of something, that it wasn't just empty space between objects. In the 1600s, amateur scientists discovered a bizarre phenomenon: the vacuum, air that seemed actually to be composed of nothing and that behaved differently from normal air. Flames would be extinguished in a vacuum; a vacuum seal was so strong that two teams of horses could not pull it apart. In 1659, the English scientist Robert Boyle had placed a bird in a jar and sucked out the air with a vacuum pump. The bird died, as Boyle suspected it might, but curiously enough, it also froze. If a vacuum was so different from normal air that it could extinguish life, that meant there must be some invisible substance that normal air was made of. And it suggested that changing the volume or pressure of gases could change their temperature. Our knowledge expanded in the eighteenth century, as the steam engine forced engineers to figure out exactly how heat and energy are converted, inventing a whole science of thermodynamics. Tools for measuring heat and weight with increased precision were developed, along with standardized scales such as Celsius and Fahrenheit, and as is so often the case in the history of science and innovation, when you have a leap forward in the accuracy of measuring something, new possibilities emerge.

All of these building blocks were circulating through Gorrie's mind, like molecules in a gas, bouncing off each other, forming new connections. In his spare time, he started to build a refrigeration machine. It would use energy from a pump to compress air. The compression heated the air. The machine then cooled down the compressed air by running it through pipes cooled with water. When the air expanded, it pulled heat from its environment, and just like the tetrahedral bonds of hydrogen dissolving into liquid water, that heat extraction cooled the surrounding air. It could even be used to create ice.

Amazingly, Gorrie's machine worked. No longer dependent on ice shipped from a thousand miles away, Gorrie reduced his patients' fevers with home-grown cold. He applied for a patent – correctly predicting a future where artificial cold, as he wrote, 'might better serve mankind . . . Fruits, vegetables, and meats will be preserved in transit by my refrigeration system and thereby enjoyed by all!'

And yet, despite his success as an inventor, Gorrie went nowhere as a businessman. Thanks to Tudor's success, natural ice was abundant and cheap when the storms didn't disrupt trade. To make things worse, Tudor himself launched a smear campaign about Gorrie's invention – claiming the ice produced by his machine was infected with bacteria. It was a classic case of a dominant industry disparaging a much more powerful new technology, the way the first computers with graphic interfaces were dismissed by their rivals as 'toys' and not 'serious business machines.' John Gorrie died penniless, having failed to sell a single machine.

But the idea of artificial cold didn't die with Gorrie. After thousands of years of neglect, suddenly the globe lit up with patents filed for some variation of artificial refrigeration. The idea was suddenly everywhere, not because people had stolen Gorrie's idea, but because they'd independently hit upon the same basic architecture. The conceptual building blocks were finally in place, and so the idea of creating artificially cold air was suddenly 'in the air.'

Those patents rippling across the planet are an example of one of the great curiosities in the history of innovation: what scholars now call 'multiple invention.' Inventions and scientific discoveries tend to come in clusters, where a handful of geographically dispersed investigators stumble independently onto the very same discovery. The isolated genius coming up with an idea that no one else could even dream of is actually the exception, not the rule. Most discoveries become imaginable at a very specific moment in history, after which point multiple people start to imagine them. The electric battery, the telegraph, the steam engine, and the digital music library were all independently invented by multiple individuals in the space of a few years. In the early 1920s, two Columbia University scholars surveyed the history of invention in a wonderful paper called 'Are Inventions Inevitable?' They found 148 instances of simultaneous invention, most of them occurring within the same decade. Hundreds more have since been discovered.

Refrigeration was no different: the knowledge of thermodynamics and the basic chemistry of air, combined with the economic fortunes being made in the ice trade, made artificial cold ripe for invention. One of those simultaneous inventors was the French engineer Ferdinand Carré, who independently designed a refrigeration machine that followed the same basic principles as Gorrie's. He built prototypes for his refrigeration machine in Paris, but his idea would ultimately triumph because of events unfolding across the Atlantic: a different kind of ice famine in the American South. After the Civil War broke out in 1861, the Union blockaded the southern states to cripple the Confederate economy. The Union navy stopped the flow of ice more effectively than did the storms that churned up along the Gulf Stream. Having built up an economic and cultural dependence on the ice trade, the sweltering southern states suddenly found themselves in desperate need of artificial cold.

As the war raged, shipments of smuggled goods could sometimes make it through the blockade at night to land at beaches along the Atlantic and Gulf coasts. But the smugglers weren't just carrying cargoes of gunpowder or weapons. Sometimes they carried goods that were far more novel: ice-making machines, based on Carré's design. These new devices used ammonia as a refrigerant and could churn out four hundred pounds of ice per hour. Carré's machines were smuggled all the way from France to Georgia, Louisiana, and Texas. A network of innovators tinkered with Carré's machines, improving their efficiency. A handful of commercial ice plants opened, marking the debut on the main stage of industrialization. By 1870, the southern states made more artificial ice than anywhere else in the world.

In the decades after the Civil War, artificial refrigeration exploded, and the natural-ice trade began its slow decline into obsolescence. Refrigeration became a huge industry, measured not just by the cash that changed hands but also in the sheer size of the machines: steam-powered monster machines weighing hundreds of tons, maintained by a full-time army of engineers. At the turn of the twentieth century, New York's Tribeca neighborhood – now home to some of the most expensive loft apartments in the world – was essentially a giant refrigerator, entire blocks of windowless buildings designed to chill the endless flood of produce from the nearby Washington food market.

Almost everything in the nineteenth-century story of cold was about making it bigger, more ambitious. But the next revolution in artificial cold would proceed in the exact opposite direction. Cold was about to get small: those block-long Tribeca refrigerators would soon shrink down to fit in every kitchen in America. But the smaller footprint of artificial cold would, ironically, end up triggering changes in human society that were so massive you could see them from space.

In the winter of 1916, an eccentric naturalist and entrepreneur moved his young family up to the remote tundra of Labrador. He had spent several winters there on his own, starting a fur company breeding foxes and occasionally shipping animals and reports back to the U.S. Biological Survey. Five weeks after the birth of his son, his wife and child joined him. Labrador was, to say the least, not an ideal place for a newborn. The climate was unforgiving, with temperatures regularly hitting 30 degrees below Fahrenheit, and the region was entirely bereft of modern medical facilities. The food, too, left a great deal to be desired. The bleak climate in Labrador meant that everything you ate during the winter was either frozen or preserved: other than the fish, there were no sources of fresh food. A typical meal would be what locals called 'brewis': salted cod and hard tack, which is rock solid bread, boiled up and garnished with 'scrunchions,' which were small, fried chunks of salted pork fat. Any meat or produce that had been frozen would be mushy and tasteless when thawed out.

But the naturalist was an adventurous eater, fascinated with the cuisines of different cultures. (In his journals, he recorded eating everything from rattlesnake to skunk.) And so he took up ice fishing with some of the local Inuits, carving holes in frozen lakes and casting a line for trout. With air temperatures so far below zero, a fish pulled out of the lake would freeze solid in a matter of seconds.

Unwittingly, the young naturalist had stumbled across a powerful scientific experiment as he sat down to eat with his family in Labrador. When they thawed out the frozen trout from the ice-fishing expeditions, they discovered it tasted far fresher than the usual grub. The difference was so striking that he became obsessed with trying to figure out why the frozen trout retained its flavor so much more effectively. And so Clarence Birdseye began an investigation that would ultimately put his name on packages of frozen peas and fish sticks in supermarkets around the world.

At first, Birdseye had assumed the trout had preserved its freshness simply because it had been caught more recently, but the more he studied the phenomenon, he began to think there was some other factor at work. For starters, ice-fished trout would retain its flavor for months, unlike other frozen fish. He began experimenting with frozen vegetables and discovered that produce frozen in the depths of winter somehow tasted better than produce frozen in late fall or early spring. He analyzed the food under a microscope and noticed a striking difference in the ice crystals that formed during the freezing process: the frozen produce that had lost its flavor had significantly larger crystals that seemed to be breaking down the molecular structure of the food itself.

Eventually, Birdseye hit upon a coherent explanation for the dramatic difference in taste: It was all about the speed of the freezing process. A slow freeze allowed the hydrogen bonds of ice to form larger crystalline shapes. But a freeze that happened in seconds – 'flash freezing,' as we now call it – generated much smaller crystals that did less damage to the food itself. The Inuit fishermen hadn't thought about it in terms of crystals and molecules, but they had been savoring the benefits of flash freezing for centuries by pulling live fish out of the water into shockingly cold air.

As his experiments continued, an idea began to form in Birdseye's mind: with artificial refrigeration becoming increasingly commonplace, the market for frozen food could be immense, assuming you could solve the quality problem. Like Tudor before him, Birdseye began taking notes on his experiments with cold. And like Tudor, the idea would linger in his mind for a decade before it turned into something commercially viable. It was not a sudden epiphany or lightbulb moment, but something much more leisurely, an idea taking shape piece by piece over time. It was what I like to call a 'slow hunch' – the anti-'lightbulb moment,' the idea that comes into focus over decades, not seconds.

The first inspiration for Birdseye had been the very pinnacle of freshness: a trout pulled out of a frozen lake. But the second would be the exact opposite: a commercial fishing ship's hull filled with rotting cod. After his Labrador adventure, Birdseye returned to his original home in New York and took a job with the Fisheries Association, where he saw firsthand the appalling conditions that characterized the commercial fishing business. 'The inefficiency and lack of sanitation in the distribution of whole fresh fish so disgusted me,' Birdseye would later write, 'that I set out to develop a method that would permit the removal of inedible waste from perishable foods at production points, packaging them in compact and convenient containers, and distributing them to the housewife with their intrinsic freshness intact.'

In the first decades of the twentieth century, the frozen-food business was considered to be the very bottom of the barrel. You could buy frozen fish or produce, but it was widely assumed to be inedible. (In fact, frozen food was so appalling that it was banned at New York State prisons for being below the culinary standards of the convicts.) One key problem was that the food was being frozen at relatively high temperatures, often just a few degrees below freezing. Yet scientific advances over the preceding decades had made it possible to artificially produce temperatures that were positively Labradorian. By the early 1920s, Birdseye had developed a flash-freezing process using stacked cartons of fish frozen at minus 40 degrees Fahrenheit. Inspired by the new industrial model of Henry Ford's Model T factory, he created a 'double-belt freezer' that ran the freezing process along a more efficient production line. He formed a company called General Seafood using these new production techniques. Birdseye found that just about anything he froze with this method – fruit, meat, vegetables – would be remarkably fresh after thawing.

Frozen food was still more than a decade away from becoming a staple of the American diet. (It required a critical mass of freezers – in supermarkets and home kitchens – that wouldn't fully come into being until the postwar years.) But Birdseye's experiments were so promising that in 1929, just months before the Black Friday crash, General Seafood was acquired by the Postum Cereal Company, which promptly changed its name to General Foods. Birdseye's adventures in ice fishing had made him a multimillionaire. His name endures on packages of frozen fish filets to this day.

Birdseye's frozen-food breakthrough took shape as a slow hunch, but it also emerged as a kind of collision between several very different geographic and intellectual spaces. To imagine a world of flash-frozen food, Birdseye needed to experience the challenges of feeding a family in an arctic climate surrounded by brutal cold; he needed to spend time with the Inuit fishermen; he needed to inspect the foul containers of cod-fishing trawlers in New York harbors; he needed the scientific knowledge of how to produce temperatures well below freezing; he needed the industrial knowledge of how to build a production line. Like every big idea, Birdseye's breakthrough was not a single insight, but a *network* of other ideas, packaged together in a new configuration. What made Birdseye's idea so powerful was not simply his individual genius, but the diversity of places and forms of expertise that he brought together.

In our age of locally sourced, artisanal food production, the frozen 'TV dinners' that arose in the decades after Birdseye's discovery have fallen out of favor. But in its original incarnation, frozen food had a positive impact on health, introducing more nutrition into the diets of Americans. Flash-frozen food extended the reach of the food network in both time and space: produce harvested in summer could be consumed months later; fish caught in the North Atlantic could be eaten in Denver or Dallas. It was better to eat frozen peas in January than it was to wait five months for fresh ones.

*

By the 1950s, Americans had adopted a lifestyle that was profoundly shaped by artificial cold, buying frozen dinners purchased in the refrigerated aisles of the local supermarket, and stacking them up in the deep freeze of their new Frigidaires, featuring the latest in ice-making technology. Behind the scenes, the entire economy of cold was supported by a vast fleet of refrigerated trucks, transporting Birds Eye frozen peas (and their many imitations) around the country.

In that iconic 1950s American household, the most novel cold-producing device was not storing fish filets for dinner or making ice for the martinis. It was cooling down (and dehumidifying) the entire house. The first 'apparatus for treating air' had been dreamed up by a young engineer named Willis Carrier in 1902. The story of Carrier's invention is a classic in the annals of accidental discovery. As a twenty-five-year-old engineer, Carrier had been hired by a printing company in Brooklyn to devise a scheme that would help them keep the ink from smearing in the humid summer months. Carrier's invention not only removed the humidity from the printing room; it also chilled the air. Carrier noticed that everyone suddenly wanted to have lunch next to the printing presses, and he began to design contraptions that would be deliberately built to regulate the humidity and temperature in an interior space. Within a few years, Carrier had formed a company – still one of the largest air-conditioning manufacturers in the world – that focused on industrial uses for the technology. But Carrier was convinced that air-conditioning should also belong to the masses.

His first great test came over Memorial Day weekend of 1925, when Carrier debuted an experimental AC system in

Paramount Pictures' new flagship Manhattan movie theater, the Rivoli. Theaters had long been oppressive places to visit during the summer months. (In fact, a number of Manhattan playhouses had experimented with ice-based cooling during the nineteenth century, with predictably moist results.) Before AC, the whole idea of a summer blockbuster would have seemed preposterous: the last place you'd want to be on a warm day was a room filled with a thousand other perspiring bodies. And so Carrier had persuaded Adolph Zukor, the legendary chief of Paramount, that there was money to be made by investing in central air for his theaters.

Zukor himself showed up for the Memorial Day weekend test, sitting inconspicuously in the balcony seats. Carrier and his team had some technical difficulties getting the AC up and running; the room was filled with hand fans waving furiously before the picture started. Carrier later recalled the scene in his memoirs:

> It takes time to pull down the temperature in a quickly filled theater on a hot day, and a still longer time for a packed house. Gradually, almost imperceptibly, the fans dropped into laps as the effects of the air conditioning system became evident. Only a few chronic fanners persisted, but soon they, too, ceased fanning . . . We then went into the lobby and waited for Mr Zukor to come downstairs. When he saw us, he did not wait for us to ask his opinion. He said tersely, 'Yes, the people are going to like it.'

Between 1925 and 1950, most Americans experienced air-conditioning only in large commercial spaces such as movie theaters, department stores, hotels, or office buildings. Carrier knew that AC was headed for the domestic sphere, but the machines were simply too large and expensive for a middle-class home. The Carrier Corporation did offer a glimpse of this future in its 1939 World's Fair attraction, 'The Igloo of Tomorrow.' In a bizarre structure that looked something like a five-story helping of soft-serve vanilla ice cream, Carrier showcased the wonders of domestic air-conditioning, accompanied by a squadron of Rockettes-style 'snow bunnies.'

But Carrier's vision of domestic cool would be postponed by the outbreak of World War II. It wasn't until the late 1940s, after almost fifty years of experimentation, that air-conditioning finally made its way to the home front, with the first in-window portable units appearing on the market. Within half a decade, Americans were installing more than a million units a year. When we think about twentieth-century miniaturization, our minds naturally gravitate to the transistor or the microchip, but the shrinking footprint of air-conditioning deserves its place in the annals of innovation as well: a machine that had once been larger than a flatbed truck reduced in size so that it could fit in a window.

That shrinking would quickly set off an extraordinary chain of events, in many ways rivaling the impact of the automobile on settlement patterns in the United States. Places that had been intolerably hot and humid – including some of the cities where Frederic Tudor had sweated out the summer as a young man – were suddenly tolerable to a much larger slice of the general public. By 1964, the historic flow of people from South to North that had characterized the post–Civil War era had been reversed. The Sun Belt expanded with new immigrants from colder states, who could put up with the tropical humidity or blazing desert climates thanks to domestic air-conditioning. Tucson rocketed from 45,000 people to 210,000 in just ten years; Houston expanded from 600,000 to 940,000 in the same decade. In the 1920s, when Willis Carrier was first demonstrating air-conditioning to Adolph Zukor at the Rivoli Theatre, Florida's population stood at less than one million. Half a century later, the state was well on the way to becoming one of the four most populous in the country, with ten million people escaping the humid summer months in air-conditioned homes. Carrier's invention circulated more than just molecules of oxygen and water. It ended up circulating *people* as well.

Broad changes in demography invariably have political effects. The migration to the Sun Belt changed the political map of America. Once a Democratic stronghold, the South was besieged by a massive influx of retirees who were more conservative in their political outlook. As the historian Nelson W. Polsby demonstrates in his book *How Congress Evolves*, Northern Republicans moving south in the post-AC era did as much to undo the 'Dixiecrat' base as the rebellion against the civil rights movement. In Congress, this had the paradoxical effect of unleashing a wave of liberal reforms, as the congressional Democrats were no longer divided between conservative Southerners and progressives in the North. But air-conditioning arguably had the most significant impact on Presidential politics. Swelling populations in Florida, Texas, and Southern California shifted the electoral college toward the Sun Belt, with warm-climate states gaining twenty-nine electoral college votes between 1940 and 1980, while the colder states of the Northeast and Rust Belt lost thirty-one. In the first half of the twentieth century, only two presidents or vice presidents hailed from Sun Belt states. Starting in 1952, however, every single winning presidential ticket contained a Sun Belt candidate, until Barack Obama and Joe Biden broke the streak in 2008.

This is long-zoom history: almost a century after Willis Carrier began thinking about keeping the ink from smearing in Brooklyn, our ability to manipulate tiny molecules of air and moisture helped transform the geography of American politics. But the rise of the Sun Belt in the United States was just a dress rehearsal for what is now happening on a planetary scale. All around the world, the fastest growing megacities are predominantly in tropical climates: Chennai, Bangkok, Manila, Jakarta, Karachi, Lagos, Dubai, Rio de Janeiro. Demographers predict that these hot cities will have more than a billion new residents by 2025.

It goes without saying that many of these new immigrants don't have air-conditioning in their homes, at least not yet, and it is an open question whether these cities are sustainable in the long run, particularly those based in desert climates. But the ability to control temperature and humidity in office buildings, stores, and wealthier homes allowed these urban centers to attract an economic base that has catapulted them to megacity status. It's no accident that the world's largest cities – London, Paris, New York, Tokyo – were almost exclusively in temperate climates until the second half of the twentieth century. What we are seeing now is arguably the largest mass migration in human history, and the first to be triggered by a home appliance.

The dreamers and inventors who ushered in the cold revolution didn't have eureka moments, and their brilliant ideas rarely transformed the world immediately. Mostly they had hunches, but they were tenacious enough to keep those hunches alive for years, even decades, until the pieces came together. Some of those innovations can seem trivial to us today. All that collective ingenuity, focused over decades and decades – all to make the world safe for the TV dinner? But the frozen world that Tudor and Birdseye helped conjure into being would do more than just populate the world with fish sticks. It would also populate the world with *people*, thanks to the flash freezing and cryopreservation of human semen, eggs, and embryos. Millions of human beings around the world owe their existence to the technologies of artificial cold. Today, new techniques in oocyte cryopreservation are allowing women to store healthy eggs in their younger years, extending their fertility well into their forties and fifties in many cases. So much of the new freedom in the way we have children now – from lesbian couples or single mothers using sperm banks to conceive, to women giving themselves two decades in the workforce before thinking about kids – would have been impossible without the invention of flash freezing.

When we think about breakthrough ideas, we tend to be constrained by the scale of the original invention. We figure out a way to make artificial cold, and we assume that will just mean that our rooms will be cooler, we'll sleep better on hot nights, or there will be a reliable supply of ice cubes for our sodas. That much is easy to understand. But if you tell the story of cold only in that way, you miss the epic scope of it. Just two centuries after Frederic Tudor started thinking about shipping ice to Savannah, our mastery of cold is helping to reorganize settlement patterns all over the planet and bring millions of new babies into the world. Ice seems at first glance like a trivial advance: a luxury item, not a necessity. Yet over the past two centuries its impact has been staggering, when you look at it from the long-zoom perspective: from the transformed landscape of the Great Plains; to the new lives and lifestyles brought into being via frozen embryos; all the way to vast cities blooming in the desert.

CHAPTER 3

Sound

Roughly one million years ago, the seas retreated from the basin that surrounds modern-day Paris, leaving a ring of limestone deposits that had once been active coral reefs. Over time, the River Cure in Burgundy slowly carved its way through some of those limestone blocks, creating a network of caves and tunnels that would eventually be festooned with stalactites and stalagmites formed by rainwater and carbon dioxide. Archeological findings suggest that Neanderthals and early modern humans used the caves for shelter and ceremony for tens of thousands of years. In the early 1990s, an immense collection of ancient paintings was discovered on the walls of the cave complex in Arcy-sur-Cure: over a hundred images of bison, woolly mammoths, birds, fish – even, most hauntingly, the imprint of a child's hand. Radiometric dating determined that the images were thirty thousand years old. Only the paintings at Chauvet, in southern France, are believed to be older.

For understandable reasons, cave paintings are conventionally cited as evidence of the primordial drive to represent the world in images. Eons before the invention of cinema, our ancestors would gather together in the firelit caverns and stare at flickering images on the wall. But in recent years, a new theory has emerged about the primitive use of the Burgundy caves, one focused not on the images of these underground passages, but rather on the *sounds*.

A few years after the paintings in Arcy-sur-Cure were discovered, a music ethnographer from the University of Paris named Iegor Reznikoff began studying the caves the way a bat would: by listening to the echoes and reverberations created in different parts of the cave complex. It had long been apparent that the Neanderthal images were clustered in specific parts of the cave, with some of the most ornate and dense images appearing more than a kilometer deep. Reznikoff determined that the paintings were consistently placed at the most acoustically interesting parts of the cave, the places where the reverberation was the most profound. If you make a loud sound standing beneath the images of Paleolithic animals at the far end of the Arcy-sur-Cure caves, you hear seven distinct echoes of your voice. The reverberation takes almost five seconds to die down after your vocal chords stop vibrating. Acoustically, the effect is not unlike the famous 'wall of sound' technique that Phil Spector used on the 1960s records he produced for artists such as the Ronettes and Ike and Tina Turner. In Spector's system, recorded sound was routed through a basement room filled with speakers and microphones that created a massive artificial echo. In Arcy-sur-Cure, the effect comes courtesy of the natural environment of the cave itself.

Reznikoff's theory is that Neanderthal communities gathered beside the images they had painted, and they chanted or sang in some kind of shamanic ritual, using the reverberations of the cave to magically widen the sound of their voices. (Reznikoff also discovered small red dots painted at other sonically rich parts of the cave.) Our ancestors couldn't record the sounds they experienced the way they recorded their visual experience of the world in paintings. But if Reznikoff is correct, those early humans were experimenting with a primitive form of sound engineering – amplifying and enhancing that most intoxicating of sounds: the human voice.

The drive to enhance – and, ultimately, reproduce – the human voice would in time pave the way for a series of social and technological breakthroughs: in communications and computation, politics and the arts. We readily accept the idea that science and technology have enhanced our vision to a remarkable extent: from spectacles to the Keck telescopes. But our vocal chords, vibrating in speech and in song, have also been massively augmented by artificial means. Our voices grew louder; they began traveling across wires laid on the ocean floor; they slipped the surly bonds of Earth and began bouncing off satellites. The essential revolutions in vision largely unfolded between the Renaissance and the Enlightenment: spectacles, microscopes, telescopes; seeing clearly, seeing very far, and seeing very close. The technologies of the voice did not arrive in full force until the late nineteenth century. When they did, they changed just about everything. But they didn't begin with amplification. The first great breakthrough in our obsession with the human voice arrived in the simple act of writing it down.

*

For thousands of years after those Neanderthal singers gathered in the reverberant sections of the Burgundy caves, the idea of recording sound was as fanciful as counting fairies. Yes, over that period we refined the art of designing acoustic spaces to amplify our voices and our instruments: medieval cathedral design, after all, was as much about sound engineering as it was about creating epic visual experiences. But no one even bothered to imagine capturing sound directly. Sound was ethereal, not tangible. The best you could do was imitate sound with your own voice and instruments.

The dream of recording the human voice entered the adjacent possible only after two key developments: one from physics, the other from anatomy. From about 1500 on, scientists began to work under the assumption that sound traveled through the air in invisible waves. (Shortly thereafter they discovered that these waves traveled four times faster through water, a curious fact that wouldn't turn out to be useful for another four centuries.) By the time of the Enlightenment, detailed books of anatomy had mapped the basic structure of the human ear, documenting the way sound waves were funneled through the auditory canal, triggering vibrations in the eardrum. In the 1850s, a Parisian printer named Édouard-Léon Scott de Martinville happened to stumble across one of these anatomy books, triggering a hobbyist's interest in the biology and physics of sound.

Scott had also been a student of shorthand writing; he'd published a book on the history of stenography a few years before he began thinking about sound. At the time, stenography was the most advanced form of voice-recording technology in existence; no system could capture the spoken word with as much accuracy and speed as a trained stenographer. But as he looked at these detailed illustrations of the inner ear, a new thought began to take shape in Scott's mind: perhaps the process of transcribing the human voice could be automated. Instead of a human writing down words, a machine could write sound waves.

In March 1857, two decades before Thomas Edison would invent the phonograph, the French patent office awarded Scott a patent for a machine that recorded sound. Scott's contraption funneled sound waves through a hornlike apparatus that ended with a membrane of parchment. Sound waves would trigger vibrations in the parchment, which would then be transmitted to a stylus made of pig's bristle. The stylus would etch out the waves on a page darkened by the carbon of lampblack. He called his invention the 'phonautograph': the self-writing of sound.

In the annals of invention, there may be no more curious mix of farsightedness and myopia than the story of the phonautograph. On the one hand, Scott had managed to make a critical conceptual leap – that sound waves could be pulled out of the air and etched onto a recording medium – more than a decade before other inventors and scientists got around to it. (When you're two decades ahead of Edison, you can be pretty sure you're doing well for yourself.) But Scott's invention was hamstrung by one crucial – even comical – limitation. He had invented the first sound recording device in history. But he forgot to include *playback*.

Actually, 'forgot' is too strong a word. It seems obvious to us now that a device for recording sound should also include a feature where you can actually *hear* the recording. Inventing the phonautograph without including playback seems a bit like inventing the automobile but forgetting to include the bit where the wheels rotate. But that is because we are judging Scott's work from the other side of the divide. The idea that machines could convey sound waves that had originated elsewhere was not at all an intuitive one; it wasn't until Alexander Graham Bell began reproducing sound waves at the end of a telephone that playback became an obvious leap. In a sense, Scott had to look around two significant blind spots, the idea that sound could be recorded *and* that those recordings could be converted back into sound waves. Scott managed to grasp the first, but he couldn't make it all the way to the second. It wasn't so much that he forgot or failed to make playback work; it was that the idea never even occurred to him.

If playback was never part of Scott's plan, it is fair to ask exactly why he bothered to build the phonautograph in the first place. What good is a record player that doesn't play records? Here we confront the double-edged sword of relying on governing metaphors, of borrowing ideas from other fields and applying them in a new context. Scott got to the idea of recording audio through the metaphor of stenography: write waves instead of words. That structuring metaphor enabled him to make the first leap, years ahead of his peers, but it also may have prevented him from making the second. Once words have been converted into the code of shorthand, the information captured there is decoded by a reader who understands the code. Scott thought the same would happen with his phonautograph. The machine would etch waveforms into the lampblack, each twitch of the stylus corresponding to some phoneme uttered by a human voice. And humans would learn to 'read' those squiggles the way they had learned to read the squiggles of shorthand. In a sense, Scott wasn't trying to invent an audio-recording device at all. He was trying to invent the ultimate transcription service – only you had to learn a whole new language in order to read the transcript.

It wasn't that crazy an idea, looking back on it. Humans had proven to be unusually good at learning to recognize visual patterns; we internalize our alphabets so well we don't even have to think about reading once we've learned how to do it. Why would sound waves, once you could get them on the page, be any different?

Sadly, the neural toolkit of human beings doesn't seem to include the capacity for reading sound waves by sight. A hundred and fifty years have passed since Scott's invention, and we have mastered the art and science of sound to a degree that would have astonished Scott. But not a single person among us has learned to visually parse the spoken words embedded in printed sound waves. It was a brilliant gamble, but ultimately a losing one. If we were going to decode recorded audio, we needed to convert it back to sound so we could do our decoding via the eardrum, not the retina.

We may not be waveform readers, but we're not exactly slackers, either: during the century and a half that followed Scott's invention, we did manage to invent a machine that could 'read' the visual image of a waveform and convert it back into sound: namely, computers. Just a few years ago, a team of sound historians named David Giovannoni, Patrick

Feaster, Meagan Hennessey, and Richard Martin discovered a trove of Scott's phonautographs in the Academy of Sciences in Paris, including one from April 1860 that had been marvelously preserved. Giovannoni and his colleagues scanned the faint, erratic lines that had been first scratched into the lampblack when Lincoln was still alive. They converted that image into a digital waveform, then played it back through computer speakers.

At first, they thought they were hearing a woman's voice, singing the French folk song '*Au clair de la lune*,' but later they realized they had been playing back the audio at double its recorded speed. When they dropped it down to the right tempo, a man's voice appeared out of the crackle and hiss: Édouard-Léon Scott de Martinville warbling from the grave.

Understandably, the recording was not of the highest quality, even played at the right speed. For most of the clip, the random noise of the recording apparatus overwhelms Scott's voice. But even this apparent failing underscores the historic importance of the recording. The strange hisses and decay of degraded audio signals would become commonplace to the twentieth-century ear. But these are not sounds that occur in nature. Sound waves dampen and echo and compress in natural environments. But they don't break up into the chaos of mechanical noise. The sound of static is a modern sound. Scott captured it first, even if it took a century and a half to hear it.

But Scott's blind spot would not prove to be a complete dead end. Fifteen years after his patent, another inventor was experimenting with the phonautograph, modifying Scott's original design to include an actual ear from a cadaver in order to understand the acoustics better. Through his tinkering, he hit upon a method for both capturing *and* transmitting sound. His name was Alexander Graham Bell.

For some reason, sound technology seems to induce a strange sort of deafness among its most advanced pioneers. Some new tool comes along to share or transmit sound in a new way, and again and again its inventor has a hard time imagining how the tool will eventually be used. When Thomas Edison completed Scott's original project and invented the phonograph in 1877, he imagined it would regularly be used as a means of sending audio letters through the postal system. Individuals would record their missives on the phonograph's wax scrolls, and then pop them into the mail, to be played back days later. Bell, in inventing the telephone, made what was effectively a mirror-image miscalculation: He envisioned one of the primary uses for the telephone to be as a medium for sharing live music. An orchestra or singer would sit on one end of the line, and listeners would sit back and enjoy the sound through the telephone speaker on the other. So, the two legendary inventors had it exactly reversed: people ended up using the phonograph to listen to music and using the telephone to communicate with friends.

As a form of media, the telephone most resembled the one-to-one networks of the postal service. In the age of mass media that would follow, new communications platforms would be inevitably drawn toward the model of big-media creators and a passive audience of consumers. The telephone system would be the one model for more intimate – one-to-one, not one-to-many – communications until e-mail arrived a hundred years later. The telephone's consequences were immense and multifarious. International calls brought the world closer together, though the threads connecting us were thin until recently. The first transatlantic line that enabled ordinary citizens to call between North America and Europe was laid only in 1956. In the first configuration, the system allowed twenty-four simultaneous calls. That was the total bandwidth for a voice conversation between the two continents just fifty years ago: out of several hundred million voices, only two dozen conversations at a time. Interestingly, the most famous phone in the world – the 'red phone' that provided a hotline between the White House and the Kremlin – was not a phone at all in its original incarnation. Created after the communications fiasco that almost brought us to nuclear war in the Cuban Missile Crisis, the hotline was actually a Teletype that enabled quick, secure messages to be sent between the powers. Voice calls were considered to be too risky, given the difficulties of real-time translation.

The telephone enabled less obvious transformations as well. It popularized the modern meaning of the word 'hello' – as a greeting that starts a conversation – transforming it into one of the most recognized words anywhere on earth. Telephone switchboards became one of the first inroads for women into the 'professional' classes. (AT&T alone employed 250,000 women by the mid-forties.) An AT&T executive named John J. Carty argued in 1908 that the telephone had had as big of an impact on the building of skyscrapers as the elevator:

> It may sound ridiculous to say that Bell and his successors were the fathers of modern commercial architecture – of the skyscraper. But wait a minute. Take the Singer Building, the Flatiron Building, the Broad Exchange, the Trinity, or any of the giant office buildings. How many messages do you suppose go in and out of those buildings every day? Suppose there was no telephone and every message had to be carried by a personal messenger? How much room do you think the necessary elevators would leave for offices? Such structures would be an economic impossibility.

But perhaps the most significant legacy of the telephone lay in a strange and marvelous organization that grew out of it: Bell Labs, an organization that would play a critical role in creating almost every major technology of the twentieth century. Radios, vacuum tubes, transistors, televisions, solar cells, coaxial cables, laser beams, microprocessors, computers, cell phones, fiber optics – all these essential tools of modern life descend from ideas originally generated at Bell Labs. Not for nothing was it known as 'the idea factory.' The interesting question about Bell Labs is not *what* it invented. (The answer to that is simple: just about everything.) The real question is *why* Bell Labs was able to create so much of the twentieth century. The definitive history of Bell Labs, Jon Gertner's *The Idea Factory*, reveals the secret to the labs' unrivaled success. It was not just the diversity of talent, and the tolerance of failure, and the willingness to make big bets – all of which were traits that Bell Labs shared with Edison's famous lab at Menlo Park as well as other research labs around the world. What made Bell Labs fundamentally different had as much to do with antitrust law as the geniuses it attracted.

From as early as 1913, AT&T had been battling the U. S. government over its monopoly control of the nation's phone service. That it was, in fact, a monopoly was undeniable. If you were making a phone call in the United States at any point between 1930 and 1984, you were almost without exception using AT&T's network. That monopoly power made the company immensely profitable, since it faced no significant competition. But for seventy years, AT&T managed to keep the regulators at bay by convincing them that the phone network was a 'natural monopoly' and a necessary one. Analog phone circuits were simply too complicated to be run by a hodgepodge of competing firms; if Americans wanted to have a reliable phone network, it needed to be run by a single company. Eventually, the antitrust lawyers in the Justice Department worked out an intriguing compromise, settled officially in 1956. AT&T would be allowed to maintain its monopoly over phone service, but any patented invention that had originated in Bell Labs would have to be freely licensed to any American company that found it useful, and all new patents would have to be licensed for a modest fee. Effectively, the government said to AT&T that it could keep its profits, but it would have to give away its ideas in return.

It was a unique arrangement, one we are not likely to see again. The monopoly power gave the company a trust fund for research that was practically infinite, but every interesting idea that came out of that research could be immediately adopted by other firms. So much of the American success in postwar electronics – from transistors to computers to cell phones – ultimately dates back to that 1956 agreement. Thanks to the antitrust resolution, Bell Labs became one of the strangest hybrids in the history of capitalism: a vast profit machine generating new ideas that were, for all practical purposes, socialized. Americans had to pay a tithe to AT&T for their phone service, but the new innovations AT&T generated belonged to everyone.

One of the most transformative breakthroughs in the history of Bell Labs emerged in the years leading up to the 1956 agreement. For understandable reasons, it received almost no attention at the time; the revolution it would ultimately enable was almost half a century in the future, and its very existence was a state secret, almost as closely guarded as the Manhattan Project. But it was a milestone nonetheless, and once again, it began with the sound of the human voice.

The innovation that had created Bell Labs in the first place – Bell's telephone – had ushered us across a crucial threshold in the history of technology: for the first time, some component of the physical world had been represented in electrical energy in a direct way. (The telegraph had converted man-made symbols into electricity, but sound belonged to nature as well as culture.) Someone spoke into a receiver, generating sound waves that became pulses of electricity that became sound waves again on the other end. Sound, in a way, was the first of our senses to be electrified. (Electricity helped us *see* the world more clearly thanks to the lightbulb during the same period, but it wouldn't record or transmit what we saw for decades.) And once those sound waves became electric, they could travel vast distances at astonishing speeds.

But as magical as those electrical signals were, they were not infallible. Traveling from city to city over copper wires, they were vulnerable to decay, signal loss, noise. Amplifiers, as we will see, helped combat the problem, boosting signals as they traveled down the line. But the ultimate goal was a pure signal, some kind of perfect representation of the voice that wouldn't degrade as it wound its way through the telephone network. Interestingly, the path that ultimately led to that goal began with a different objective: not keeping our voices pure, but keeping them *secret.*

During World War II, the legendary mathematician Alan Turing and Bell Labs' A. B. Clark collaborated on a secure communications line, code-named SIGSALY, that converted the sound waves of human speech into mathematical expressions. SIGSALY recorded the sound wave twenty thousand times a second, capturing the amplitude and frequency of the wave at that moment. But that recording was not done by converting the wave into an electrical signal or a groove in a wax cylinder. Instead, it turned the information into numbers, encoded it in the binary language of zeroes and ones. 'Recording,' in fact, was the wrong word for it. Using a term that would become common parlance among hip-hop and electronic musicians fifty years later, they called this process 'sampling.' Effectively, they were taking snapshots of the sound wave twenty thousand times a second, only those snapshots were written out in zeroes and ones: digital, not analog.

Working with digital samples made it much easier to

transmit them securely: anyone looking for a traditional analog signal would just hear a blast of digital noise. (SIGSALY was code-named 'Green Hornet' because the raw information sounded like a buzzing insect.) Digital signals could also be mathematically encrypted much more effectively than analog signals. While the Germans intercepted and recorded many hours of SIGSALY transmissions, they were never able to interpret them.

Developed by a special division of the Army Signal Corps, and overseen by Bell Labs researchers, SIGSALY went into operation on July 15, 1943, with a historic transatlantic phone call between the Pentagon and London. At the outset of the call, before the conversation turned to the more pressing issues of military strategy, the president of Bell Labs, Dr O. E. Buckley, offered some introductory remarks on the technological breakthrough that SIGSALY represented:

> We are assembled today in Washington and London to open a new service, secret telephony. It is an event of importance in the conduct of the war that others here can appraise better than I. As a technical achievement, I should like to point out that it must be counted among the major advances in the art of telephony. Not only does it represent the achievement of a goal long sought – complete secrecy in radiotelephone transmission – but it represents the first practical application of new methods of telephone transmission that promise to have far-reaching effects.

If anything, Buckley underestimated the significance of those 'new methods.' SIGSALY was not just a milestone in telephony. It was a watershed moment in the history of media and communications more generally: for the first time, our experiences were being digitized. The technology behind SIGSALY would continue to be useful in supplying secure lines of communication. But the truly disruptive force that it unleashed would come from another strange and wonderful property it possessed: digital copies could be perfect copies. With the right equipment, digital samples of sound could be transmitted and copied with perfect fidelity. So much of the turbulence of the modern media landscape – the reinvention of the music business that began with file-sharing services such as Napster, the rise of streaming media, and the breakdown of traditional television networks – dates back to the digital buzz of the Green Hornet. If the robot historians of the future had to mark one moment where the 'digital age' began – the computational equivalent of the Fourth of July or Bastille Day – that transatlantic phone call in July 1943 would certainly rank high on the list. Once again, our drive to reproduce the sound of the human voice had expanded the adjacent possible. For the first time, our experience of the world was becoming digital.

The digital samples of SIGSALY traveled across the Atlantic courtesy of another communications breakthrough that Bell Labs helped create: radio. Interestingly, while radio would eventually become a medium saturated with the sound of people talking or singing, it did not begin that way. The first functioning radio transmissions – created by Guglielmo Marconi and a number of other more-or-less simultaneous inventors in the last decades of the nineteenth century – were almost exclusively devoted to sending Morse code.

(Marconi called his invention 'wireless telegraphy.') But once information began flowing through the airwaves, it was not long before the tinkerers and research labs began thinking of how to make spoken word and song part of the mix.

One of those tinkerers was Lee De Forest, one of the most brilliant and erratic inventors of the twentieth century. Working out of his home lab in Chicago, De Forest dreamed of combining Marconi's wireless telegraph with Bell's telephone. He began a series of experiments with a spark-gap transmitter, a device that created a bright, monotone pulse of electromagnetic energy that can be detected by antennae miles away, perfect for sending Morse code. One night, while De Forest was triggering a series of pulses, he noticed something strange happening across the room: every time he created a spark, the flame in his gas lamp turned white and increased in size. Somehow, De Forest thought, the electromagnetic pulse was intensifying the flame. That flickering gaslight planted a seed in De Forest's head: somehow a gas could be used to amplify weak radio reception, perhaps making it strong enough to carry the more information-rich signal of spoken words and not just the staccato pulses of Morse code. He would later write, with typical grandiosity: 'I discovered an Invisible Empire of the Air, intangible, yet solid as granite.'

After a few years of trial and error, De Forest settled on a gas-filled bulb containing three precisely configured electrodes designed to amplify incoming wireless signals. He called it the Audion. As a transmission device for the spoken word, the Audion was just powerful enough to transmit intelligible signals. In 1910, De Forest used an Audion-equipped radio device to make the first ever ship-to-shore broadcast of the human voice. But De Forest had much more ambitious plans for his device. He had imagined a world in which his wireless technology was used not just for military and business communications but also for mass enjoyment – and in particular, to make his great passion, opera, available to everyone. 'I look forward to the day when opera may be brought into every home,' he told the *New York Times*, adding, somewhat less romantically, 'Someday even advertising will be sent out over the wireless.'

On January 13, 1910, during a performance of *Tosca* by New York's Metropolitan Opera, De Forest hooked up a telephone microphone in the hall to a transmitter on the roof to create the first live public radio broadcast. Arguably the most poetic of modern inventors, De Forest would later describe his vision for the broadcast: 'The ether wave passing over the tallest towers and those who stand between are unaware of the silent voices which pass them on either side . . . And when it speaks to him, the strains of some well-loved earthly melody, his wonder grows.'

Alas, this first broadcast did not trigger quite as much wonder as it did derision. De Forest invited hordes of reporters and VIPs to listen to the broadcast on his radio receivers dotted around the city. The signal strength was terrible, and listeners heard something closer to the Green Hornet's unintelligible buzz than the strains of a well-loved earthly melody. The *Times* declared the whole adventure 'a disaster.' De Forest was even sued by the U. S. attorney for fraud, accused of overselling the value of the Audion in wireless technology, and briefly incarcerated. Needing cash to pay his legal bills, De Forest sold the Audion patent at a bargain price to AT&T.

When the researchers at Bell Labs began investigating the Audion, they discovered something extraordinary: from the very beginning Lee De Forest had been flat-out wrong about most of what he was inventing. The increase in the gas flame had nothing to do with electromagnetic radiation. It was caused by sound waves from the loud noise of the spark. Gas didn't detect and amplify a radio signal at all; in fact, it made the device less effective.

But somehow, lurking behind all of De Forest's accumulation of errors, a beautiful idea was waiting to emerge. Over the next decade, engineers at Bell Labs and elsewhere modified his basic three-electrode design, removing the gas from the bulb so that it sealed a perfect vacuum, transforming it into both a transmitter *and* a receiver. The result was the vacuum tube, the first great breakthrough of the electronics revolution, a device that would boost the electrical signal of just about any technology that needed it. Television, radar, sound recording, guitar amplifiers, X-rays, microwave ovens, the 'secret telephony' of SIGSALY, the first digital computers – all would rely on vacuum tubes. But the first mainstream technology to bring the vacuum tube into the home was radio. In a way, it was the realization of De Forest's dream: an empire of air transmitting well-loved melodies into living rooms everywhere. And yet, once again, De Forest's vision would be frustrated by actual events. The melodies that started playing through those magical devices were well-loved by just about everyone except De Forest himself.

*

Radio began its life as a two-way medium, a practice that continues to this day as ham radio: individual hobbyists talking to one another over the airwaves, occasionally eavesdropping on other conversations. But by the early 1920s, the broadcast model that would come to dominate the technology had evolved. Professional stations began delivering packaged news and entertainment to consumers who listened on radio receivers in their homes. Almost immediately, something entirely unexpected happened: the existence of a mass medium for sound unleashed a new kind of music on the United States, a music that had until then belonged almost exclusively to New Orleans, to the river towns of the American South, and to African-American neighborhoods in New York and Chicago. Almost overnight, radio made jazz a national phenomenon. Musicians such as Duke Ellington and Louis Armstrong became household names. Ellington's band performed weekly national broadcasts from the Cotton Club in Harlem starting in the late 1920s; Armstrong became the first African-American to host his own national radio show shortly thereafter.

All of this horrified Lee De Forest, who wrote a characteristically baroque denunciation to the National Association of Broadcasters: 'What have you done with my child, the radio broadcast? You have debased this child, dressed him in rags of ragtime, tatters of jive and boogie-woogie.' In fact, the technology that De Forest had helped invent was intrinsically better suited to jazz than it was to classical performances. Jazz punched through the compressed, tinny sound of early AM radio speakers; the vast dynamic range of a symphony was largely lost in translation. The blast of Satchmo's trumpet played better on the radio than the subtleties of Schubert.

The collision of jazz and radio created, in effect, the first surge of a series of cultural waves that would wash over twentieth-century society. A new sound that has been slowly incubating in some small section of the world – New Orleans, in the case of jazz – finds its way onto the mass medium of radio, offending the grown-ups and electrifying the kids. The channel first carved out by jazz would subsequently be filled by rock 'n' roll from Memphis, British pop from Liverpool, rap and hip-hop from South Central and Brooklyn. Something about radio and music seems to have encouraged this pattern, in a way that television or film did not: almost immediately after a national medium emerged for sharing music, subcultures of sound began flourishing on that medium. There were 'underground' artists before radio – impoverished poets and painters – but radio helped create a template that would become commonplace: the underground artist who becomes an overnight celebrity.

With jazz, of course, there was a crucial additional element. The overnight celebrities were, for the most part, African-Americans: Ellington, Armstrong, Ella Fitzgerald, Billie Holiday. It was a profound breakthrough: for the first time, white America welcomed African-American culture into its living room, albeit through the speakers of an AM radio. The jazz stars gave white America an example of African-Americans becoming famous and wealthy and admired for their skills as entertainers rather than advocates. Of course, many of those musicians also became powerful advocates, in songs such as Billie Holiday's 'Strange Fruit,' with its bitter tale of a southern lynching. Radio signals had a kind of freedom to them that proved to be liberating in the real world. Those radio waves ignored the way in which society was segmented at that time: between black and white worlds, between different economic classes. The radio signals were color-blind. Like the Internet, they didn't break down barriers as much as live in a world separate from them.

The birth of the civil rights movement was intimately bound up in the spread of jazz music throughout the United States. It was, for many Americans, the first cultural common ground between black and white America that had been largely created by African-Americans. That in itself was a great blow to segregation. Martin Luther King Jr made the connection explicit in remarks he delivered at the Berlin Jazz Festival in 1964:

> It is no wonder that so much of the search for identity among American Negroes was championed by Jazz musicians. Long before the modern essayists and scholars wrote of 'racial identity' as a problem for a multi-racial world, musicians were returning to their roots to affirm that which was stirring within their souls. Much of the power of our Freedom Movement in the United States has come from this music. It has strengthened us with its sweet rhythms when courage began to fail. It has calmed us with its rich harmonies when spirits were down. And now, Jazz is exported to the world.

Like many political figures of the twentieth century, King was indebted to the vacuum tube for another reason. Shortly after De Forest and Bell Labs began using vacuum tubes to enable radio broadcasts, the technology was enlisted to amplify the human voice in more immediate settings: powering amplifiers attached to microphones, allowing people to speak or sing to massive crowds for the first time in history. Tube amplifiers finally allowed us to break free from the sound engineering that had prevailed since Neolithic times. We were no longer dependent on the reverberations of caves or cathedrals or opera houses to make our voices louder. Now electricity could do the work of echoes, but a thousand times more powerfully.

Amplification created an entirely new kind of political event: mass rallies oriented around individual speakers. Crowds had played a dominant role in political upheaval for the preceding century and a half; if there is an iconic image of revolution before the twentieth century, it's the swarm of humanity taking the city streets in 1789 or 1848. But amplification took those teeming crowds and gave them a focal point: the voice of the leader reverberating through the plaza or stadium or park. Before tube amplifiers, the limits of our vocal chords made it difficult to speak to more than a thousand people at a time. (The elaborate vocal stylings of opera singing were in many ways designed to coax maximum projection out of the biological limitations of the voice.) But a microphone attached to multiple speakers extended the range of earshot by several orders of magnitude. No one recognized – and exploited – this new power more quickly than Adolf Hitler, whose Nuremberg rallies addressed more than a hundred thousand followers, all fixated on the amplified sound of the Führer's voice. Remove the microphone and amplifier from the toolbox of twentieth-century technology and you remove one of that century's defining forms of political organization, from Nuremberg to 'I Have a Dream.'

Tube amplification enabled the musical equivalent of political rallies as well: the Beatles playing Shea Stadium, Woodstock, Live Aid. But the idiosyncrasies of vacuum-tube technology also had a more subtle effect on twentieth-century music – making it not just loud but also making it noisy.

It is hard for those of us who have lived in the postindustrial world our entire lives to understand just how much a shock the sound of industrialization was to human ears a century or two ago. An entirely new symphony of discord suddenly entered the realm of everyday life, particularly in large cities: the crashing, clanging of metal on metal; the white-noise blast of the steam engine. The noise was, in many ways, as shocking as the crowds and the smells of the big cities. By the 1920s, as electrically amplified sounds began roaring alongside the rest of the urban tumult, organizations such as Manhattan's Noise Abatement Society began advocating for a quieter metropolis. Sympathetic to the society's mission, a Bell Labs engineer named Harvey Fletcher created a truck loaded with state-of-the-art sound equipment and Bell engineers who drove slowly around New York City noise hot spots taking sound measurements. (The unit of measurement for sound volume – the decibel – came out of Fletcher's research.) Fletcher and his team found that some city sounds – riveting and drilling in construction, the roar of the subway – were at the decibel threshold for auditory pain. At Cortlandt Street, known as 'Radio Row,' the noise of storefronts showcasing the latest radio speakers was so loud it even drowned out the elevated train.

But while noise-abatement groups battled modern noise through regulations and public campaigns, another reaction emerged. Instead of being repelled by the sound, our ears began to find something beautiful in it. The routine experiences of everyday life had been effectively a training session for the aesthetics of noise since the early nineteenth century. But it was the vacuum tube that finally brought noise to the masses.

Starting in the 1950s, guitarists playing through tube amplifiers noticed that they could make an intriguing new kind of sound by overdriving the amp: a crunchy layer of noise on top of the notes generated by strumming the strings of the guitar itself. This was, technically speaking, the sound of the amplifier malfunctioning, distorting the sound it had been designed to reproduce. To most ears it sounded like something was broken with the equipment, but a small group of musicians began to hear something appealing in the sound. A handful of early rock 'n' roll recordings in the 1950s features a modest amount of distortion on the guitar tracks, but the art of noise wouldn't really take off until the sixties. In July 1960, a bassist named Grady Martin was recording a riff for a Marty Robbins song called 'Don't Worry' when his amplifier malfunctioned, creating a heavily distorted sound that we now call a 'fuzz tone.' Initially Robbins wanted it removed from the song, but the producer persuaded him to keep it. 'No one could figure out the sound because it sounded like a saxophone,' Robbins would say years later. 'It sounded like a jet engine taking off. It had many different sounds.' Inspired by the strange, unplaceable noise of Martin's riff, another band called the Ventures asked a friend to hack together a device that could add the fuzz effect deliberately. Within a year, there were commercial distortion boxes on the market; within three years Keith Richards was saturating the opening riff of 'Satisfaction' with distortion, and the trademark sound of the sixties was born.

A similar pattern developed with a novel – and initially unpleasant – sound that occurs when amplified speakers and microphones share the same physical space: the swirling, screeching noise of feedback. Distortion was a sound that had at least some aural similarity to those industrial sounds that had first emerged in the eighteenth century. (Hence the 'jet engine' tone of Grady Martin's bass line.) But feedback was an entirely new creature; it did not exist in any form until the invention of speakers and microphones roughly a century ago. Sound engineers would go to great lengths to eliminate feedback from recordings or concert settings, positioning microphones so they didn't pick up signal from the speakers, and thus cause the infinite-loop screech of feedback. Yet once again, one man's malfunction turned out to be another man's music, as artists such as Jimi Hendrix or Led Zeppelin – and later punk experimentalists like Sonic Youth – embraced the sound in their recordings and performances. In a real sense, Hendrix was not just playing the guitar on those feedback-soaked recordings in the late 1960s, he was creating a new sound that drew upon the vibration of the guitar strings, the microphone-like pickups on the guitar itself, and the speakers, building on the complex and unpredictable interactions between those three technologies.

Sometimes cultural innovations come from using new technologies in unexpected ways. De Forest and Bell Labs weren't trying to invent the mass rally when they came up with the first sketches of a vacuum tube, but it turned out to be easy to assemble mass rallies once you had amplification to share a single voice with that many people. But sometimes the innovation comes from a less likely approach: by deliberately exploiting the malfunctions, turning noise and error into a useful signal. Every genuinely new technology has a genuinely new way of breaking – and every now and then, those malfunctions open a new door in the adjacent possible. In the case of the vacuum tube, it trained our ears to enjoy a sound that would no doubt have made Lee De Forest recoil in horror. Sometimes the way a new technology breaks is almost as interesting as the way it works.

From the Neanderthals chanting in the Burgundy caves, to Édouard-Léon Scott de Martinville warbling into his phonautograph, to Duke Ellington broadcasting from the Cotton Club, the story of sound technology had always been about extending the range and intensity of our voices and our ears. But the most surprising twist of all would come just a century ago, when humans first realized that sound could be harnessed for something else: to help us see.

The use of light to signal the presence of dangerous shorelines to sailors is an ancient practice; the Lighthouse of Alexandria, constructed several centuries before the birth of Christ, was one of the original seven wonders of the world. But lighthouses perform poorly at precisely the point where they are needed the most: in stormy weather, where the light they transmit is obscured by fog and rain. Many lighthouses employed warning bells as an additional signal, but those too could be easily drowned out by the sound of a roaring sea. Yet sound waves turn out to have an intriguing physical property: under water, they travel four times faster than they do through the air, and they are largely undisturbed by the sonic chaos above sea level.

In 1901, a Boston-based firm called the Submarine Signal Company began manufacturing a system of communications tools that exploited this property of aquatic sound waves: underwater bells that chimed at regular intervals, and microphones specially designed for underwater reception called 'hydrophones.' The SSC established more than a hundred stations around the world at particularly treacherous harbors or channels, where the underwater bells would warn vessels, equipped with the company's hydrophones, that steered too close to the rocks or shoals. It was an ingenious system, but it had its limits. To begin with, it worked only in places where the SSC had installed warning bells. And it was entirely useless at detecting less predictable dangers: other ships, or icebergs.

The threat posed by icebergs to maritime travel became vividly apparent to the world in April 1912, when the *Titanic* foundered in the North Atlantic. Just a few days before the sinking, the Canadian inventor Reginald Fessenden had run across an engineer from the SSC at a train station, and after a quick chat, the two men agreed that Fessenden should come by the office to see the latest underwater signaling technologies. Fessenden had been a pioneer of wireless radio, responsible for both the first radio transmission of human speech and the first transatlantic two-way radio transmission of Morse code. That expertise had led the SSC to ask him to help them design their hydrophone system to better filter out the background noise of underwater acoustics. When news of the *Titanic* broke, just four days after his visit to the SSC, Fessenden was as shocked as the rest of the world, but unlike the rest of the world, he had an idea about how to prevent these tragedies in the future.

Fessenden's first suggestion had been to replace the bells with a continuous, electric-powered tone that could also be used to transmit Morse code, borrowing from his experiences with wireless telegraphy. But as he tinkered with the possibilities, he realized the system could be much more ambitious. Instead of merely listening to sounds generated by specifically designed and installed warning posts, Fessenden's device would generate its *own* sounds onboard the ship and listen to the echoes created as those new sounds bounced off objects in the water, much as dolphins use echolocation to navigate their way around the ocean. Borrowing the same principles that had attracted the cave chanters to the unusually reverberant sections of the Arcy-sur-Cure caves, Fessenden tuned the device so that it would resonate with only a small section of the frequency spectrum, right around 540hz, allowing it to ignore all the background noise of the aquatic environment. After calling it, somewhat disturbingly, his 'vibrator' for a few months, he ultimately dubbed it the 'Fessenden Oscillator.' It was a system for both sending and receiving underwater telegraphy, and the world's first functional sonar device.

Once again, the timing of world-historical events underscored the need for Fessenden's contraption. Just a year after he completed his first working prototype, World War I erupted. The German U-boats roaming the North Atlantic now posed an even greater threat to maritime travel than the *Titanic*'s iceberg. The threat was particularly acute for Fessenden, who as a Canadian citizen was a fervent patriot of the British Empire. (He also seems to have been a borderline racist, later advancing a theory in his memoirs about why 'blond-haired men of English extraction' had been so central to modern innovation.) But the United States was still two years away from joining the war, and the executives at the SSC didn't share his allegiance to the Union Jack. Faced with the financial risk of developing *two* revolutionary new technologies, the company decided to build and market the oscillator as a wireless telegraphy device exclusively.

Fessenden ultimately traveled on his own dollar all the way to Portsmouth, England, to try to persuade the Royal Navy to invest in his oscillator, but they too were dubious of this miracle invention. Fessenden would later write: 'I pleaded with them to just let us open the box and show them what the apparatus was like.' But his pleas were ultimately ignored. Sonar would not become a standard component of naval warfare until World War II. By the armistice in 1918, upward of ten thousand lives had been lost to the U-boats. The British and, eventually, the Americans had experimented with countless offensive and defensive measures to ward off these submarine predators. But, ironically, the most valuable defensive weapon would have been a simple 540hz sound wave, bouncing off the hull of the attacker.

In the second half of the twentieth century, the principles of echolocation would be employed to do far more than detect icebergs and submarines. Fishing vessels – and amateur fishers – used variations of Fessenden's oscillator to detect their catch. Scientists used sonar to explore the last great mysteries of our oceans, revealing hidden landscapes, natural resources, and fault lines. Eighty years after the sinking of the *Titanic* inspired Reginald Fessenden to dream up the first sonar, a team of American and French researchers used sonar to discover the vessel on the Atlantic seabed, twelve thousand feet below the surface.

But Fessenden's innovation had the most transformative effect on dry land, where ultrasound devices, using sound to see into a mother's womb, revolutionized prenatal care, allowing today's babies and their mothers to be routinely saved from complications that had been fatal less than a century ago. Fessenden had hoped his idea – using sound to see – might save lives; while he couldn't persuade the authorities to put it to use in detecting U-boats, the oscillator did end up saving millions of lives, both at sea and in a place Fessenden would never have expected: the hospital.

Of course, ultrasound's most familiar use involves determining the sex of a baby early in a pregnancy. We are accustomed now to think of information in binary terms: a zero or a one, a circuit connected or broken. But in all of life's experiences, there are few binary crossroads like the sex of your unborn child. Are you going to have a girl or a boy? How many life-changing consequences flow out of that simple unit of information? Like many of us, my wife and I learned the gender of our children using ultrasound. We now have other, more accurate, means of determining the sex of a fetus, but we found our way to that knowledge first by bouncing sound waves off the growing body of our unborn child. Like the Neanderthals navigating the caves of Arcy-sur-Cure, echoes led the way.

There is, however, a dark side to that innovation. The introduction of ultrasound in countries such as China with a strong cultural preference for male offspring has led to a growing practice of sex-selective abortions. An extensive supply of ultrasound machines was introduced throughout China in the early 1980s, and while the government shortly thereafter officially banned the use of ultrasound to determine sex, the 'back-door' use of the technology for sex selection is widespread. By the end of the decade, the sex ratio at birth in hospitals throughout China was almost 110 boys to every 100 girls, with some provinces reporting ratios as high as 118:100. This may be one of the most astonishing, and tragic, hummingbird effects in all of twentieth-century technology: someone builds a machine to listen to sound waves bouncing off icebergs, and a few generations later, millions of female fetuses are aborted thanks to that very same technology.

The skewed sex ratios of modern China contain several important lessons, setting aside the question of abortion itself, much less gender-based abortion. First, they are a reminder that no technological advance is purely positive in its effects: for every ship saved from an iceberg, there are countless pregnancies terminated because of a missing Y chromosome. The march of technology has its own internal logic, but the moral application of that technology is up to us. We can decide to use ultrasound to save lives or terminate them. (Even more challenging, we can use ultrasound to blur the very boundaries of life, detecting a heartbeat in a fetus that is only weeks old.) For the most part, the adjacencies of technological and scientific progress dictate what we can invent next. However smart you might be, you can't invent an ultrasound before the discovery of sound waves. But what we decide to do with those inventions? That is a more complicated question, one that requires a different set of skills to answer.

But there's another, more hopeful lesson in the story of sonar and ultrasound, which is how quickly our ingenuity is able to leap boundaries of conventional influence. Our ancestors first noticed the power of echo and reverberation to change the sonic properties of the human voice tens of thousands of years ago; for centuries we have used those properties to enhance the range and power of our vocal chords, from cathedrals to the Wall of Sound. But it's hard to imagine anyone studying the physics of sound two hundred years ago predicting that those echoes would be used to track undersea weapons or determine the sex of an unborn child. What began with the most moving and intuitive sound to the human ear – the sound of our voices in song, in laughter, sharing news or gossip – has been transformed into the tools of both war and peace, death and life. Like those distorted wails of the tube amp, it is not always a happy sound. Yet, again and again, it turns out to have unsuspected resonance.

CHAPTER 4

Clean

In December 1856, a middle-aged Chicago engineer named Ellis Chesbrough traveled across the Atlantic to take in the monuments of the European continent. He visited London, Paris, Hamburg, Amsterdam, and a half dozen other towns – the classic Grand Tour. Only Chesbrough hadn't made his pilgrimage to study the architecture of the Louvre or Big Ben. He was there, instead, to study the invisible achievements of European engineering. He was there to study the sewers.

Chicago, in the middle of the nineteenth century, was a city in dire need of expertise about waste removal. Thanks to its growing role as a transit hub bringing wheat and preserved pork from the Great Plains to the coastal cities, the city had gone from hamlet to metropolis in a matter of decades. But unlike other cities that had grown at prodigious rates during this period (such as New York and London), Chicago had one crippling attribute, the legacy of a glacier's crawl thousands of years before the first humans settled there: it was unforgivingly flat. During the Pleistocene era, vast ice fields crept down from Greenland, covering present-day Chicago with glaciers that were more than a mile high. As the ice melted, it formed a massive body of water that geologists now call Lake Chicago. As that lake slowly drained down to form Lake Michigan, it flattened the clay deposits left behind by the glacier. Most cities enjoy a reliable descending grade down to the rivers or harbors they evolved around. Chicago, by comparison, is an ironing board – appropriately enough, for the great city of the American plains.

Building a city on perfectly flat land would seem like a good problem to have; you would think hilly, mountainous terrain like that of San Francisco, Cape Town, or Rio would pose more engineering problems, for buildings and for transportation. But flat topographies don't drain. And in the middle of the nineteenth century, gravity-based drainage was key to urban sewer systems. Chicago's terrain also suffered from being unusually nonporous; with nowhere for the water to go, heavy summer rainstorms could turn the topsoil into a murky marshland in a matter of minutes. When William Butler Ogden, who would later become Chicago's inaugural mayor, first waded through the rain-soaked town, he found himself 'sinking knee deep in the mud.' He wrote to his brother-in-law, who had purchased land in the frontier town in a bold bet on its future potential: 'You have been guilty of an act of great folly in making [this] purchase.' In the late 1840s, roadways made out of wood planks had been erected over the mud; one contemporary noted that every now and then one of the planks would give way, and 'green and black slime [would] gush up between the cracks.' The primary system for sanitation removal was scavenging pigs roaming the streets, devouring the refuse that the humans left behind.

With its rail and shipping network expanding at extraordinary speed, Chicago more than tripled in size during the 1850s. That rate of growth posed challenges for the city's housing and transportation resources, but the biggest strain of all came from something more scatological: when almost a hundred thousand new residents arrive in your city, they generate a lot of excrement. One local editorial declared: 'The gutters are running with filth at which the very swine turn up their noses in supreme disgust.' We rarely think about it, but the growth and vitality of cities have always been dependent on our ability to manage the flow of human waste that emerges when people crowd together. From the very beginnings of human settlements, figuring out where to put all the excrement has been just as important as figuring out how to build shelter or town squares or marketplaces.

The problem is particularly acute in cities experiencing runaway growth, as we see today in the favelas and shantytowns of megacities. Nineteenth-century Chicago, of course, had both human and animal waste to deal with, the horses in the streets, the pigs and cattle awaiting slaughter in the stockyards. ('The river is positively red with blood under the Rush Street Bridge and past down our factory,' one industrialist wrote. 'What pestilence may result from it I don't know.') The effects of all this filth were not just offensive to the senses; they were deadly. Epidemics of cholera and dysentery erupted regularly in the 1850s. Sixty people died a day during the outbreak of cholera in the summer of 1854. The authorities at the time didn't fully understand the connection between waste and disease. Many of them subscribed to the then-prevailing 'miasma' theory, contending that epidemic disease arose from poisonous vapors, sometimes called 'death fogs,' that people inhaled in dense cities. The true transmission route – invisible bacteria carried in fecal matter polluting the water supply – would not become conventional wisdom for another decade.

But while their bacteriology wasn't well developed, the Chicago authorities were right to make the essential connection between cleaning up the city and fighting disease. On February 14, 1855, a Chicago Board of Sewerage Commissioners was created to address the problem; their first act was to announce a search for 'the most competent engineer of the time who was available for the position of chief engineer.' Within a few months, they had found their man, Ellis Chesbrough, the son of a railway officer who had worked on canal and rail projects, and who was currently employed as chief engineer of the Boston Water Works.

It was a wise choice: Chesbrough's background in railway and canal engineering turned out to be decisive in solving the problem of Chicago's flat, nonporous terrain. Creating an artificial grade by building sewers deep underground was deemed too expensive: tunneling that far below the surface was difficult work using nineteenth-century equipment, and the whole scheme required pumping the waste back to the surface at the end of the process. But here Chesbrough's unique history helped him come up with an alternate scenario, reminding him of a tool he had seen as a young man working the railway: the jackscrew, a device used to lift multiton locomotives onto the tracks. If you couldn't dig down to create a proper grade for drainage, why not use jackscrews to lift the city up?

Aided by the young George Pullman, who would later

make a fortune building railway cars, Chesbrough launched one of the most ambitious engineering projects of the nineteenth century. Building by building, Chicago was lifted by an army of men with jackscrews. As the jackscrews raised the buildings inch by inch, workmen would dig holes under the building foundations and install thick timbers to support them, while masons scrambled to build a new footing under the structure. Sewer lines were inserted beneath buildings with main lines running down the center of streets, which were then buried in landfill that had been dredged out of the Chicago River, raising the entire city almost ten feet on average. Tourists walking around downtown Chicago today regularly marvel at the engineering prowess on display in the city's spectacular skyline; what they don't realize is that the ground beneath their feet is also the product of brilliant engineering. (Not surprisingly, having participated in such a Herculean undertaking, when George Pullman set out to build his model factory town of Pullman, Illinois, several decades later, his first step was to install sewer and water lines before breaking ground on any of the buildings.)

Amazingly, life went on largely undisturbed as Chesbrough's team raised the city's buildings. One British visitor observed a 750-ton hotel being lifted, and described the surreal experience in a letter: 'The people were in [the hotel] all the time coming and going, eating and sleeping – the whole business of the hotel proceeding without interruption.' As the project advanced, Chesbrough and his team became ever more daring in the structures they attempted to raise. In 1860, engineers raised half a city block: almost an acre of five-story buildings weighing an estimated thirty-five thousand tons was lifted by more than six thousand jackscrews. Other structures had to be moved as well as lifted to make way for the sewers: 'Never a day passed during my stay in the city,' one visitor recalled, 'that I did not meet one or more houses shifting their quarters. One day I met nine. Going out on Great Madison Street in the horse cars we had to stop twice to let houses get across.'

The result was the first comprehensive sewer system in any American city. Within three decades, more than twenty cities around the country followed Chicago's lead, planning and installing their own underground networks of sewer tunnels. These massive underground engineering projects created a template that would come to define the twentieth-century metropolis: the idea of a city as a system supported by an invisible network of subterranean services. The first steam train traveled through underground tunnels beneath London in 1863. The Paris metro opened in 1900 followed shortly by the New York subway. Pedestrian walkways, automobile freeways, electrical and fiber-optic cabling coiled their way beneath city streets. Today, entire parallel worlds exist underground, powering and supporting the cities that rise above them. We think of cities intuitively now in terms of skylines, that epic reach toward the heavens. But the grandeur of those urban cathedrals would be impossible without the hidden world below grade.

Of all those achievements, more than the underground trains and high-speed Internet cables, the most essential and the most easily overlooked is the small miracle that sewer systems in part make possible: enjoying a glass of clean drinking water from a tap. Just a hundred and fifty years ago, in cities around the world, drinking water was effectively playing Russian roulette. When we think of the defining killers of nineteenth-century urbanism, our minds naturally turn to Jack the Ripper haunting the streets of London. But the real killers of the Victorian city were the diseases bred by contaminated water supplies.

This was the – literally – fatal flaw in Chesbrough's plan for the sewers of Chicago. He had brilliantly conceived a strategy to get the waste away from the streets and the privies and the cellars of everyday life, but almost all of his sewer pipes drained into the Chicago River, which emptied directly into Lake Michigan, the primary source of the city's drinking water. By the early 1870s, the city's water supply was so appalling that a sink or tub would regularly be filled with dead fish, poisoned by the human filth and then hoovered up into the city's water pipes. In summer months, according to one observer, the fish 'came out cooked and one's bathtub was apt to be filled with what squeamish citizens called chowder.'

Upton Sinclair's novel *The Jungle* is generally considered to be the most influential literary work in the muckraking tradition of political activism. Part of the power of the book came from its literal muckraking, describing the filth of turn-of-the-century Chicago in excruciating detail, as in this description of the wonderfully named Bubbly Creek, an offshoot of the Chicago River:

> The grease and chemicals that are poured into it undergo all sorts of strange transformations, which are the cause of its name; it is constantly in motion, as if huge fish were feeding

> in it, or great leviathans disporting themselves in its depths. Bubbles of carbonic gas will rise to the surface and burst, and make rings two or three feet wide. Here and there the grease and filth have caked solid, and the creek looks like a bed of lava; chickens walk about on it, feeding, and many times an unwary stranger has started to stroll across, and vanished temporarily.

Chicago's experience was replicated around the world: sewers removed human waste from people's basements and backyards, but more often than not they simply poured it into the drinking water supply, either directly, as in the case of Chicago, or indirectly during heavy rainstorms. Drawing plans for sewer lines and water pipes on the scale of the city itself would not be sufficient for the task of keeping the big city clean and healthy. We also needed to understand what was happening on the scale of microorganisms. We needed both a germ theory of disease – and a way to keep those germs from harming us.

When you go back to look at the initial reaction from the medical community to the germ theory, the response seems beyond comical; it simply doesn't compute. It is a well-known story that the Hungarian physician Ignaz Semmelweis was roundly mocked and criticized by the medical establishment when he first proposed, in 1847, that doctors and surgeons wash their hands before attending to their patients. (It took almost half a century for basic antiseptic behaviors to take hold among the medical community, well after Semmelweis himself lost his job and died in an insane asylum.) Less commonly known is that Semmelweis based his initial argument on studies of puerperal (or 'childbed') fever, where new mothers died shortly after childbirth. Working in Vienna's General Hospital, Semmelweis stumbled across an alarming natural experiment: the hospital contained two maternity wards, one for the well-to-do, attended by physicians and medical students, the other for the working class who received their care from midwives. For some reason, the death rates from puerperal fever were much lower in the working-class ward. After investigating both environments, Semmelweis discovered that the elite physicians and students were switching back and forth between delivering babies and doing research with cadavers in the morgue. Clearly some kind of infectious agent was being transmitted from the corpses to the new mothers; with a simple application of a disinfectant such as chlorinated lime, the cycle of infection could be stopped in its tracks.

There may be no more startling example of how much things have changed in our understanding of cleanliness over the past century and a half: Semmelweis was derided and dismissed not just for daring to propose that doctors wash their hands; he was derided and dismissed for proposing that doctors wash their hands *if they wanted to deliver babies and dissect corpses in the same afternoon.*

This is one of those places where our basic sensibilities deviate from the sensibilities of our nineteenth-century ancestors. They look and act like modern people in many ways: they take trains and schedule meetings and eat in restaurants. But every now and then, strange gaps open between us and them, not just the obvious gaps in technological sophistication, but more subtle, conceptual gaps. In today's world, we think of hygiene in fundamentally different ways. The concept of bathing, for instance, was alien to most nineteenth-century Europeans and Americans. You might naturally assume that taking a bath was a foreign concept simply because people didn't have access to running water and indoor plumbing and showers the way most of us in the developed world do today. But, in fact, the story is much more complicated than that. In Europe, starting in the Middle Ages and running almost all the way to the twentieth century, the prevailing wisdom on hygiene maintained that submerging the body in water was a distinctly unhealthy, even dangerous thing. Clogging one's pores with dirt and oil allegedly protected you from disease. 'Bathing fills the head with vapors,' a French doctor advised in 1655. 'It is the enemy of the nerves and ligaments, which it loosens, in such a way that many a man never suffers from gout except after bathing.'

You can see the force of this prejudice most clearly in the accounts of royalty during the 1600s and 1700s – in other words, the very people who could afford to have baths constructed and drawn for them without a second thought. Elizabeth I bothered to take a bath only once a month, and she was a veritable clean freak compared to her peers. As a child, Louis XIII was not bathed once until he was seven years old. Sitting naked in a pool of water was simply not something civilized Europeans did; it belonged to the barbaric traditions of Middle Eastern bathhouses, not the aristocracy of Paris or London.

Slowly, starting in the early nineteenth century, the attitudes began to shift, most notably in England and America.

Charles Dickens built an elaborate cold-water shower in his London home, and was a great advocate for the energizing and hygienic virtues of a daily shower. A minor genre of self-help books and pamphlets emerged, teaching people how to take a bath, with detailed instructions that seem today as if they are training someone to land a 747. One of the first steps Professor Higgins takes in reforming Eliza Doolittle in George Bernard Shaw's *Pygmalion* is getting her into a tub. ('You expect me to get into that and wet myself all over?' she protests. 'Not me. I should catch my death.') Harriet Beecher Stowe and her sister Catharine Beecher advocated a daily wash in their influential handbook, *The American Woman's Home*, published in 1869. Reformers began building public baths and showers in urban slums around the country. 'By the last decades of the century,' the historian Katherine Ashenburg writes, 'cleanliness had become firmly linked not only to godliness but also to the American way.'

The virtues of washing oneself were not self-evident, the way we think of them today. They had to be discovered and promoted, largely through the vehicles of social reform and word of mouth. Interestingly, there is very little discussion of soap in the popular embrace of bathing in the nineteenth century. It was hard enough just to convince people that the water wasn't going to kill them. (As we will see, when soap finally hit the mainstream in the twentieth century, it would be propelled by another new convention: advertising.) But the evangelists for bathing were supported by the convergence of several important scientific and technological developments. Advances of public infrastructure meant that people were much more likely to have running water in their homes to fill their bathtubs; that the water was cleaner than it had been a few decades earlier; and, most important, that the germ theory of disease had gone from fringe idea to scientific consensus.

This new paradigm had been achieved through two parallel investigations. First, there was the epidemiological detective work of John Snow in London, who first proved that cholera was caused by contaminated water and not miasmatic smells, by mapping the deaths of a Soho epidemic. Snow never managed to see the bacteria that caused cholera directly; the technology of microscopy at the time made it almost impossible to see organisms (Snow called them 'animalcules') that were so small. But he was able to detect the organisms indirectly, in the patterns of death on the streets of London. Snow's waterborne theory of disease would ultimately deliver the first decisive blow to the miasma paradigm, though Snow himself didn't live to see his theory triumph. After his untimely death in 1858, *The Lancet* ran a terse obituary that made no reference whatsoever to his groundbreaking epidemiological work. In 2014, the publication ran a somewhat belated 'correction' to the obit, detailing the London doctor's seminal contributions to public health.

The modern synthesis that would come to replace the miasma hypothesis – that diseases such as cholera and typhoid are caused not by smells but by invisible organisms that thrive in contaminated water – was ultimately dependent, once again, on an innovation in glass. The German lens crafters Zeiss Optical Works began producing new microscopes in the early 1870s – devices that for the first time had been constructed around mathematical formulas that described the

1. Pectoral in gold cloissoné with semiprecious stones and glass paste, with a winged scarab, symbol of resurrection, in centre, from the tomb of pharaoh Tutankhamun.

2. Circa 1900: Roman civilization, first-second century AD. Glass containers for ointments.

3. Earliest image of monk with glasses, 1342.

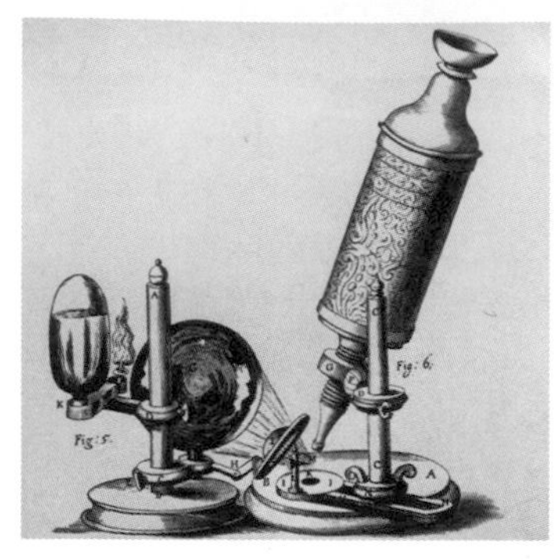

4. An early microscope designed by Robert Hooke, 1665.

5. Charles Vernon Boys standing in a laboratory, 1917.

6. Keck Observatory, Mauna Kea, Hawaii.

7. *Las Meninas* by Diego Rodriguez de Silva y Velázquez.

27. The swinging altar lamp inside Duomo Di Pisa.

28. Galileo Galilei.

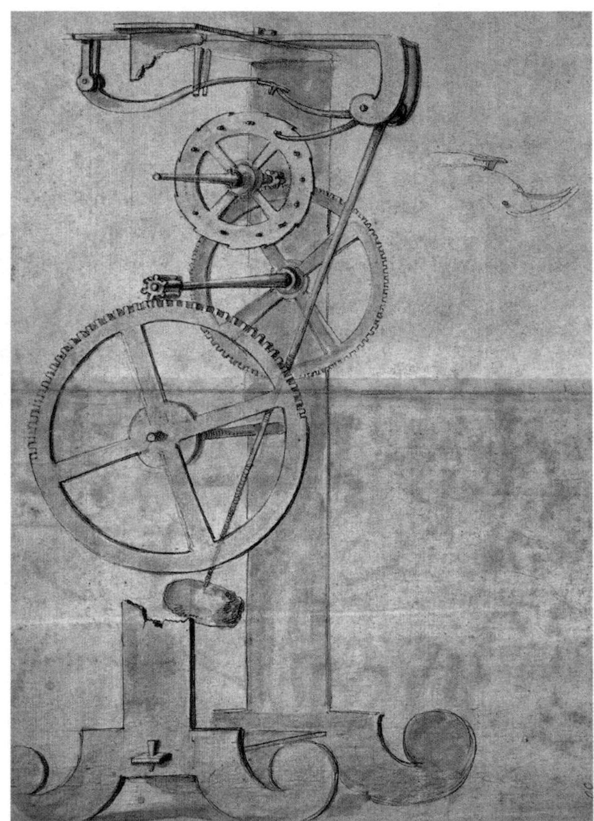

29. Drawing of the pendulum clock designed by Italian physicist, mathematician, astronomer, and philosopher Galileo Galilei in 1638–1659.

30. Aaron Dennison.

31. The rewinding of a big Dennison watch (operation done once a year) in Holborn, London.

32. Professor Charles H. Townes, executive of the physics department at Columbia University, is shown with the atomic clock in the university's physics department. Date released: January 25, 1955.

33. Chalice-shaped lamp from the tomb of Tutankhamun. The cup was intended to be filled with oil and when the wick was lit, then the scene of Tutankhamun and Ankhesenamun was visible. Ancient Egyptian. New Kingdom, Eighteenth Dynasty, 1333–1323 BC.

34. Thomas Edison.

35. New York: *Adapting the Brush Electric Light to the Illumination Of The Streets, A scene near The Fifth Avenue Hotel.*

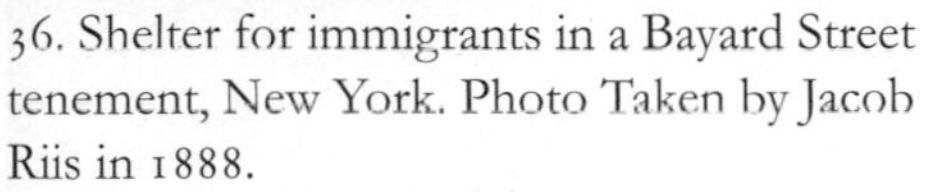

36. Shelter for immigrants in a Bayard Street tenement, New York. Photo Taken by Jacob Riis in 1888.

37. 1960s Night Scene Downtown Las Vegas, Nevada.

38. Vaughn Draggoo inspects a huge target chamber at the National Ignition Facility in California, a future test site for light-induced nuclear fusion. Beams from 192 lasers will be aimed at a pellet of fusion fuel to produce a controlled thermonuclear blast.

39. Augusta Ada King, Countess Lovelace circa 1840.

40. Charles Babbage.

41. Babbage's analytical engine.

behavior of light. These new lenses enabled the microbiological work of scientists such as Robert Koch, one of the first scientists to identify the cholera bacterium. (After receiving the Nobel Prize for his work in 1905, Koch wrote to Carl Zeiss, 'A large part of my success I owe to your excellent microscopes.') With his great rival Louis Pasteur, Koch and his microscopes helped develop and evangelize the germ theory of disease. From a technological standpoint, the great nineteenth-century breakthrough in public health – the knowledge that invisible germs can kill – was a kind of team effort between maps and microscopes.

Today, Koch is rightly celebrated for the numerous microorganisms that he identified through those Zeiss lenses. But his research also led to a related breakthrough that was every bit as important, though less widely appreciated. Koch didn't just *see* the bacteria; he also developed sophisticated tools to *measure* the density of bacteria in a given quantity of water. He mixed contaminated water with transparent gelatin, and viewed the growing bacterial colonies on a glass plate. Koch established a unit of measure that could be applied to any quantity of water – below 100 colonies per milliliter was considered to be safe to drink.

New ways of measuring create new ways of making. The ability to measure bacterial content allowed a completely new set of approaches to the challenges of public health. Before the adoption of these units of measurement, you had to test improvements to the water system the old-fashioned way: you built a new sewer or reservoir or pipe, and you sat around and waited to see if fewer people would die. But being able to take a sample of water and determine empirically whether it was free of contamination meant that cycles of experimentation could be tremendously accelerated.

Microscopes and measurement quickly opened a new front in the war on germs: instead of fighting them indirectly, by routing the waste away from the drinking water, new chemicals could be used to attack the germs directly. One of the key soldiers on this second front was a New Jersey doctor named John Leal. Like John Snow before him, Leal was a doctor who treated patients but who also had a passionate interest in wider issues of public health, particularly those concerning contaminated water supplies. It was an interest born of a personal tragedy: his father had suffered a slow and painful death from drinking bacteria-infested water during the Civil War. His father's experience in the war gives us a compelling statistical portrait of the threat posed by contaminated water and other health risks during this period. Nineteen men in the 144th Regiment died in combat, while 178 died of disease during the war.

Leal experimented with many techniques for killing bacteria, but one poison in particular began to pique his interest as early as 1898: calcium hypochlorite, the potentially lethal chemical that is better known as chlorine, also known at the time as 'chloride of lime.' The chemical had already been in wide circulation as a public health remedy: houses and neighborhoods that had suffered an outbreak of typhoid or cholera were routinely disinfected with the chemical, an intervention that did nothing to combat waterborne disease. But the idea of putting chlorine in water had not yet taken hold. The sharp, acrid smell of chloride of lime was indelibly associated with epidemic disease in the minds of city dwellers throughout the United States and Europe. It was certainly not a smell that one wanted to detect in one's drinking water. Most doctors and public health authorities rejected the approach. One noted chemist protested: 'The idea itself of chemical disinfection is repellent.' But armed with tools that enabled him to both see the pathogens behind diseases such as typhoid and dysentery and measure their overall presence in the water, Leal became convinced that chlorine – at the right dosage – could rid water of dangerous bacteria more effectively than any other means, without any threat to the humans drinking it.

Eventually, Leal landed a job with the Jersey City Water Supply Company, giving him oversight of seven billion gallons of drinking water in the Passaic River watershed. This new job set the stage for one of the most bizarre and daring interventions in the history of public health. In 1908, the company was immersed in a prolonged legal battle over contracts (worth hundreds of millions of dollars in today's money) for reservoirs and water-supply pipes they had recently completed. The judge in the case had criticized the firm for not supplying waste that was 'pure and wholesome' and ordered them to construct expensive additional sewer lines designed to keep pathogens out of the city's drinking water. But Leal knew the sewer lines would be limited in their effectiveness, particularly during big storms. And so he decided to put his recent experiments with chlorine to the ultimate test.

In almost complete secrecy, without any permission from government authorities (and no notice to the general public), Leal decided to add chlorine to the Jersey City reservoirs.

With the help of engineer George Warren Fuller, Leal built and installed a 'chloride of lime feed facility' at the Boonton Reservoir outside Jersey City. It was a staggering risk, given the popular opposition to chemical filtering at the time. But the court rulings had severely limited his timeline, and he knew that lab tests would be meaningless to a lay audience. 'Leal did not have time for a pilot study. He certainly did not have time to build a demonstration-scale facility to test the new technology,' Michael J. McGuire writes in his account, *The Chlorine Revolution*. 'If the chlorine of lime feed system lost control of the amount of chemical being fed and a slug of high chlorine residual was delivered to Jersey City, Leal knew that would define the failure of the process.'

It was the first mass chlorination of a city's water supply in history. Once word got out, however, it initially seemed as though Leal was a madman or some kind terrorist. Drinking a few glasses of calcium hypochlorite could kill you, after all. But Leal had done enough experiments to know that very small quantities of the compound were harmless to humans but lethal to many forms of bacteria. Three months after his experiment, Leal was called to appear in court to defend his actions. Throughout his interrogation, he stood strong in defense of his public health innovation:

> *Q: Doctor, what other places in the world can you mention in which this experiment has been tried of putting this bleaching powder in the same way in the drinking water of a city of 200,000 inhabitants?*
>
> A: 200,000 inhabitants? There is no such place in the world, it has never been tried.

Q: It never has been.

A: Not under such conditions or under such circumstances but it will be used many times in the future, however.

Q: Jersey City is the first one?

A: The first to profit by it.

Q: Jersey City is the first one used to prove whether your experiment is good or bad?

A: No, sir, to profit by it. The experiment is over.

Q: Did you notify the city that you were going to try this experiment?

A: I did not.

Q: Do you drink this water?

A: Yes sir.

Q: Would you have any hesitation about giving it to your wife and family?

A: I believe it is the safest water in the world.

Ultimately the court case was settled with near complete victory for Leal. 'I do there find and report,' the special master in the case wrote, 'that this device is capable of rendering the water delivered to Jersey City, pure and wholesome . . . and is effective in removing from the water . . . dangerous germs.' Within a few years, the data supporting Leal's daring move had become incontrovertible: communities such as Jersey City that enjoyed chlorinated drinking water saw dramatic decreases in waterborne diseases like typhoid fever.

At one point in Leal's cross-examination during the Jersey City trial, the prosecuting attorney accused John Leal of seeking vast financial rewards from his chlorine innovation. 'And if the experiment turned out well,' he sneered, 'why, you made a fortune.' Leal interrupted him from the witness box with a shrug, 'I don't know where the fortune comes in; it is all the same to me.' Unlike others, Leal made no attempt to patent the chlorination technique that he had pioneered at the Boonton Reservoir. His idea was free to be adopted by any water company that wished to provide its customers with 'pure and wholesome' water. Unencumbered by patent restrictions and licensing fees, municipalities quickly adopted chlorination as a standard practice, across the United States and eventually around the world.

About a decade ago, two Harvard professors, David Cutler and Grant Miller, set out to ascertain the impact of chlorination (and other water filtration techniques) between 1900 and 1930, the period when they were implemented across the United States. Because extensive data existed for rates of disease and particularly infant mortality in different communities around the country, and because chlorination systems were rolled out in a staggered fashion, Cutler and Miller were able to get an extremely accurate portrait of chlorine's effect on public health. They found that clean drinking water led to a 43 percent reduction in total mortality in the average American city. Even more impressive, chlorine and filtration systems reduced infant mortality by 74 percent, and child mortality by almost as much.

It is important to pause for a second to reflect on the significance of those numbers, to take them out of the dry domain of public health statistics and into the realm of lived experience. Until the twentieth century, one of the givens of being a parent was that you faced a very high likelihood that at least one of your children would die at an early age. What may well be the most excruciating experience that we can confront – the loss of a child – was simply a routine fact of existence. Today, in the developed world at least, that routine fact has been turned into a rarity. One of the most fundamental challenges of being alive – keeping your children safe from harm – was dramatically lessened, in part through massive engineering projects, and in part through the invisible collision between compounds of calcium hypochlorite and microscopic bacteria. The people behind that revolution didn't become rich, and very few of them became famous. But they left an imprint on our lives that is in many ways more profound than the legacy of Edison or Rockefeller or Ford.

Chlorination wasn't just about saving lives, though. It was also about having fun. After World War I, ten thousand chlorinated public baths and pools opened across America; learning how to swim became a rite of passage. These new aquatic public spaces were the leading edge in challenges to the old rules of public decency during the period between the wars. Before the rise of municipal pools, women bathers generally dressed as though they were bundled up for a sleigh ride. By the mid-1920s, women began exposing their legs below the knee; one-piece suits with lower necklines emerged a few years later. Open-backed suits, followed by two-piece outfits, followed quickly in the 1930s. 'In total, a woman's thighs, hip line, shoulders, stomach, back and breast line all become publicly exposed between 1920 and 1940,' the historian Jeff Wiltse writes in his social history of swimming,

Contested Waters. We can measure the transformation in terms of simple material: at the turn of the century, the average woman's bathing suit required ten yards of fabric; by the end of the 1930s, one yard was sufficient. We tend to think of the 1960s as the period when shifting cultural attitudes led to the most dramatic change in everyday fashion, but it is hard to rival the rapid-fire unveiling of the female body that occurred between the wars. Of course, it is likely that women's fashion would have found another route to exposure without the rise of swimming pools, but it seems unlikely that it would have happened as quickly as it did. No doubt exposing the thighs of female bathers was not in the forefront of John Leal's mind as he dumped his chlorine into the Jersey City reservoir, but like the hummingbird's wing, a change in one field triggers a seemingly unrelated change at a different order of existence: a trillion bacteria die at the hands of calcium hypochlorite, and somehow, twenty years later, basic attitudes toward exposing the female body are reinvented. As with so many cultural changes, it's not that the practice of chlorination single-handedly transformed women's fashion; many social and technological forces converged to make those bathing suits smaller: various strands of early feminism, the fetishizing gaze of the Hollywood camera, not to mention individual stars who wore those more revealing suits. But without the mass adoption of swimming as a leisure activity, those fashions would have been deprived of one of their key showcases. What's more, those other explanations – as valid as they are – usually get all the press. Ask your average person on the street what factors drive women's fashion, and they'll inevitably point to Hollywood or glossy magazines. But they won't often mention calcium hypochlorite.

Through the nineteenth century, the march of clean technologies had largely unfolded on the terrain of public health: big engineering projects, mass filtration systems. But the story of hygiene in the twentieth century is a much more intimate affair. Just a few years after Leal's bold experiment, five San Francisco entrepreneurs invested a hundred dollars each to launch a chlorine-based product. It seems with hindsight to have been a good idea, but their bleach business had been aimed at big industry, and sales didn't develop as quickly as they had hoped. But the wife of one of the investors, Annie Murray, a shop owner in Oakland, California, had an idea: that chlorine bleach could be a revolutionary product for people's *homes* as well as factories. At Murray's insistence, the company created a weaker version of the chemical and packaged it in smaller bottles. Murray was so convinced of the product's promise that she gave out free samples to all her shop customers. Within months, bottles were selling like crazy. Murray didn't realize it at that time, but she was helping to invent an entirely new industry. Annie Murray had created America's first commercial bleach for the home, and the first in a wave of cleaning brands that would become ubiquitous in the new century: Clorox.

Clorox bottles became so commonplace that the remnants our grandparents left behind are used by archaeologists to date dig sites today. (The pint-glass chlorine bleach bottle is to the early twentieth century what spear tips are to the iron age or colonial pottery is to the eighteenth century.) It was accompanied by other bestselling hygiene products for the home: Palmolive soap, Listerine, a popular antiperspirant named Odorono. Hygiene products like these were among the first to be promoted in full-page advertisements in magazines and newspapers. By the 1920s, Americans were being bombarded by commercial messages convincing them that they were facing certain humiliation if they didn't do something about the germs on their bodies or in their homes. (The phrase 'often a bridesmaid, never a bride' originated with a 1925 Listerine advertisement.) When radio and television began experimenting with storytelling, it was the personal-hygiene companies that once again led the way in pioneering new forms of advertising, a brilliant marketing move that still lingers with us today in the phrase 'soap opera.' This is one of the stranger hummingbird effects of contemporary culture: the germ theory of disease may have reduced infant mortality to a fraction of its nineteenth-century levels, and made surgery and childbirth far safer than it had been in Semmelweis's day. But it also played a crucial role in inventing the modern advertising business.

Today the cleaning business is worth an estimated $80 billion. Walk into a big-box supermarket or drugstore, and you will find hundreds, if not thousands, of products devoted to ridding our households of dangerous germs: cleaning our sinks and our toilets and floors and silverware, our teeth and our feet. These stores are effectively giant munitions depots for the war on bacteria. Naturally, there are some who feel that our obsession with cleanliness may now have gone too far. Some research suggests that our ever cleaner world may actually be linked to increasing rates of asthma and allergies, as our childhood immune systems now develop without being exposed to the full diversity of germs.

The conflict between man and bacteria that played out over the past two centuries has had far-reaching consequences: from the trivial pursuits of swimwear fashion all the way to the existential improvements of lowered infant mortality rates. Our growing understanding of the microbial routes of disease enabled cities to burst through the population ceilings that had constrained them for the entirety of human civilization. As of 1800, no society had successfully built and sustained a city of more than two million people. The first cities to challenge that barrier (London and Paris, followed shortly by New York) had suffered mightily from the diseases that erupted when that many people shared such a small amount of real estate. Many reasonable observers of urban life in the middle of the nineteenth century were convinced that cities were not meant to be built on this scale, and that London would inevitably collapse back to a more manageable size, as Rome had done almost two thousand years before. But solving the problems of clean drinking water and reliable waste removal changed all of that. A hundred and fifty years after Ellis Chesbrough first took his grand tour of European sewage, cities such as London and New York were approaching ten million residents, with life expectancies and infectious disease rates that were far lower than their Victorian antecedents.

Of course, the problem now is not cities of two million or ten million; it's megacities such as Mumbai or São Paulo that will soon embrace thirty million human beings or more, many of them living in improvised communities – shantytowns, favelas – that are closer to the Chicago that Chesbrough had to raise than a contemporary city in the developed world. If you look only at today's Chicago or London, the story of the past century and a half seems to be one of incontrovertible progress: the water is cleaner; the mortality rates are much lower; epidemic disease is effectively nonexistent. And yet today there are more than three billion people around the world who lack access to clean drinking water and basic sanitation systems. In absolute numbers, we have gone backward as a species. (There were only a billion people alive in 1850.) So the question before us now is how we bring the clean revolution to the favelas, and not just Michigan Avenue. The conventional assumption has been that these communities need to follow the same path charted by Snow, Chesbrough, Leal, and all the other unsung heroes of our public health infrastructure: they need toilets connected to massive sewer systems that dispose of waste without contaminating reservoirs that pump out filtered water, delivered through an equally elaborate system direct to the home. But increasingly, these new megacities' citizens – and other global development innovators – have begun to think that history need not repeat itself.

However bold and determined John Leal was, if he had been born just a generation earlier, he would have never had the opportunity to chlorinate the Jersey City water, because the science and the technology that made chlorination possible simply hadn't been invented yet. The maps and lenses and chemistry and units of measure that converged in the second half of the nineteenth century gave him a platform for the experiment, so much so that it is probably fair to assume that if Leal hadn't brought chlorination to the mainstream, someone else would have done it within a decade, if not sooner. All of which leads to the question: If new ideas and new technology can make a new solution imaginable, the way the germ theory and the microscope triggered the idea of chemically treating water, then has there not been a sufficient supply of new ideas since Leal's day that might trigger a *new* paradigm for keeping our cities clean, one that would bypass the big-engineering phase altogether? And perhaps that paradigm might be a leading indicator of a future that we're all destined to share. The developing world has famously bypassed some of the laborious infrastructure of wired telephone lines, jumping ahead of more 'advanced' economies by basing their communications around wireless connections instead. Could the same pattern play out with sewers?

In 2011, the Bill and Melinda Gates Foundation announced a competition to help spur a paradigm shift in the way we think about basic sanitation services. Memorably called the 'Reinvent the Toilet Challenge,' the competition solicited designs for toilets that do not require a sewer connection or electricity and cost less than five cents per user per day. The winning entry was a toilet system from Caltech that uses photovoltaic cells to power an electrochemical reactor that treats human waste, producing clean water for flushing or irrigation and hydrogen that can be stored in fuel cells. The system is entirely self-contained; it has no need for an electrical grid, a sewer line, or a treatment facility. The only input the toilet requires, beyond sunlight and human waste, is simple table salt, which is oxidized to make chlorine to disinfect the water.

Those chlorine molecules might well be the only part of the toilet that John Leal would recognize, were he around to see it today. And that's because the toilet depends on new ideas and technology that have become part of the adjacent possible in the twentieth century, tools that hopefully can allow us to bypass the costly, labor-intensive work of building giant infrastructure projects. Leal needed microscopes and chemistry and the germ theory to clean the water supply in Jersey City; the Caltech toilet needs hydrogen fuel cells, solar panels, and even lightweight, inexpensive computer chips to monitor and regulate the system.

Ironically, those microprocessors are themselves, in part, the by-product of the clean revolution. Computer chips are fantastically intricate creations – despite the fact that they are ultimately the product of human intelligence, their microscopic detail is almost impossible for us to comprehend. To measure them, we need to zoom down to the scale of micrometers, or microns: one-millionth of a meter. The width of a human hair is about a hundred microns. A single cell of your skin is about thirty microns. A cholera bacterium is about three microns across. The pathways and transistors through which electricity flows on a microchip – carrying those signals that represent the zeroes and ones of binary code – can be as small as one-tenth of a single micron. Manufacturing at this scale requires extraordinary robotics and laser tools; there's no such thing as a hand-crafted microprocessor. But chip factories also require another kind of technology, one we don't normally associate with the high-tech world: they need to be preposterously clean. A spec of household dust landing on one of these delicate silicon wafers would be comparable to Mount Everest landing in the streets of Manhattan.

Environments such as the Texas Instruments microchip plant outside Austin, Texas, are some of the cleanest places on the planet. To even enter into the space, you have to don a full clean suit, your body covered head-to-toe with sterile materials that don't shed. There's something strangely inverted about the process. Normally when you find yourself dressing in such extreme protective outfits, you're guarding yourself against some kind of hostile environment: severe cold, pathogens, the vacuum of space. But in the clean room, the suit is designed to protect the space from *you*. You are the pathogen, threatening the valuable resources of computer chips waiting to be born: your hair follicles and your epidermal layers and the mucus swarming around you. From the microchip's point of view, every human being is Pig Pen, a dust cloud of filth. Washing up before entering the clean room, you're not even allowed to use soap, because most soaps have fragrances that give off potential contaminants. Even soap is too dirty for the clean room.

There is a strange symmetry to the clean room as well, one that brings us back to those first pioneers struggling to purify the drinking water of their cities: to Ellis Chesbrough, John Snow, John Leal. Producing microchips also requires large quantities of water, only this water is radically different from the water you drink from the tap. To avoid impurities, chip plants create pure H_2O, water that has been filtered not only of any bacterial contaminants but also of all the minerals, salts, and random ions that make up normal filtered water. Stripped of all those extra 'contaminants,' ultrapure water, as it is called, is the ideal solvent for microchips. But those missing elements also make ultrapure water undrinkable for humans; chug a glass of the stuff and it will start leeching minerals out of your body. This is the full circle of clean: some of the most brilliant ideas in science and engineering in the nineteenth century helped us purify water that was too dirty to drink. And now, a hundred and fifty years later, we've created water that's *too clean* to drink.

Standing in the clean room, the mind naturally drifts back to the sewers that lie beneath our city streets, the two polar extremes of the history of clean. To build the modern world, we had to create an unimaginably repellent space, an underground river of filth, and cordon it off from everyday life. And at the same time, to make the digital revolution, we had to create a hyper-clean environment, and once again, cordon it off from everyday life. We never get to visit these environments, and so they retreat from our consciousness. We celebrate the things they make possible – towering skyscrapers and ever-more-powerful computers – but we don't celebrate the sewers and the clean rooms themselves. Yet their achievements are everywhere around us.

CHAPTER 5

Time

In October 1967, a group of scientists from around the world gathered in Paris for a conference with the unassuming name 'The General Conference on Weights and Measures.' If you've had the questionable fortune to attend an academic conference before, you probably have some sense of how these affairs go: papers are presented, along with an interminable series of panel discussions, broken up by casual networking over coffee; there's gossip and infighting at the hotel bar at night; everyone has a tolerably good time, and not a whole lot gets done. But the General Conference on Weights and Measures broke from that venerable tradition. On October 13, 1967, the attendees agreed to change the very definition of time.

For almost the entire span of human history, time had been calculated by tracking the heavenly rhythms of solar bodies. Like the earth itself, our sense of time revolved around the sun. Days were defined by the cycle of sunrise and sunset, months by the cycles of the moon, years by the slow but predictable rhythms of the seasons. For most of that stretch, of course, we misunderstood what was causing those patterns, assuming that the sun was revolving around the earth, and not the reverse. Slowly, we built tools to measure the flow of time more predictably: sundials to track the passage of the day; celestial observatories such as Stonehenge to track seasonal milestones like the summer solstice. We began dividing up time into shorter units – seconds, minutes, hours – with many of those units relying on a base-12 counting system passed down from the ancient Egyptians and Sumerians. Time was defined by grade-school division: a minute was one-sixtieth of an hour, an hour was one-twenty-fourth of a day. And a day was simply the time that passed between the two moments when the sun was highest in the sky.

But starting about sixty years ago, as our tools of measuring time increased in precision, we began to notice flaws in that celestial metronome. The clockwork of the heavens turned out to be, well, a bit wobbly. And that's what the General Conference on Weights and Measures set out to address in 1967. If we were going to be truly accurate with our measurements of time, we needed to trade the largest entity in the solar system for one of the smallest.

Measured purely by tourist attention, the Duomo di Pisa is generally overshadowed by its famous leaning neighbor next door, but the thousand-year-old cathedral, with its brilliant white stone and marble façade, is in many ways a more impressive structure than the tilted bell tower beside it. Stand at the base of the nave and gaze up toward the fourteenth-century apse mosaic, and you can re-create a moment of absentminded distraction that would ultimately transform our relationship to time. Suspended from the ceiling is a collection of altar lamps. They are motionless now, but legend has it that in 1583, a nineteen-year-old student at the University of Pisa attended prayers at the cathedral and, while daydreaming in the pews, noticed one of the altar lamps swaying back and forth. While his companions dutifully recited the Nicene Creed around him, the student became almost hypnotized by the lamp's regular motion. No matter how large the arc, the lamp appeared to take the same amount of time to swing back and forth. As the arc decreased in length, the speed of the lamp decreased as well. To confirm his observations, the student measured the lamp's swing against the only reliable clock he could find: his own pulse.

Most nineteen-year-olds figure out less scientific ways to be distracted while attending mass, but this college freshman happened to be Galileo Galilei. That Galileo was daydreaming about time and rhythm shouldn't surprise us: his father was a music theorist and played the lute. In the middle of the sixteenth century, playing music would have been one of the most temporally precise activities in everyday culture. (The musical term 'tempo' comes from the Italian word for time.) But machines that could keep a reliable beat didn't exist in Galileo's age; the metronome wouldn't be invented for another few centuries. So watching the altar lamp sway back and forth with such regularity planted the seed of an idea in Galileo's young mind. As is so often the case, however, it would take decades before the seed would blossom into something useful.

Galileo spent the next twenty years becoming a professor of mathematics, experimenting with telescopes, and more or less inventing modern science, but he managed to keep the memory of that swinging altar lamp alive in his mind. Increasingly obsessed with the science of dynamics, the study of how objects move through space, he decided to build a pendulum that would re-create what he had observed in the Duomo of Pisa so many years before. He discovered that the time it takes a pendulum to swing is not dependent on the size of the arc or the mass of the object swinging, but only on the length of the string. 'The marvelous property of the pendulum,' he wrote to fellow scientist Giovanni Battista Baliani, 'is that it makes all its vibrations, large or small, in equal times.'

In equal times. In Galileo's age, any natural phenomenon or mechanical device that displayed this rhythmic precision seemed miraculous. Most Italian towns in that period had large, unwieldy mechanical clocks that kept a loose version of the correct time, but they had to be corrected by sundial readings constantly or they would lose as much as twenty minutes a day. In other words, the state of the art in time-keeping technology was challenged by just staying accurate on the scale of *days*. The idea of a timepiece that might be accurate to the *second* was preposterous.

Preposterous, and seemingly unnecessary. Just like Frederic Tudor's ice trade, it was an innovation that had no natural market. You couldn't keep accurate time in the middle of the sixteenth century, but no one really noticed, because there was no need for split-second accuracy. There were no buses to catch, or TV shows to watch, or conference calls to join. If you knew roughly what hour of the day it was, you could get by just fine.

The need for split-second accuracy would emerge not from the calendar but from the map. This was the first great age of global navigation, after all. Inspired by Columbus, ships were sailing to the Far East and the newly discovered Americas, with vast fortunes awaiting those who navigated the oceans successfully. (And almost certain death awaiting those who got lost.) But sailors lacked any way to determine longitude at sea. Latitude you could gauge just by looking up at the sky. But before modern navigation technology, the only way to figure out a ship's longitude involved two clocks. One clock was set to the exact time of your origin point (assuming you knew the longitude of that location). The other clock recorded the current time at your location at sea. The difference between the two times told you your longitudinal position: every four minutes of difference translated to one degree of longitude, or sixty-eight miles at the equator.

In clear weather, you could easily reset the ship clock through accurate readings of the sun's position. The problem was the home-port clock. With timekeeping technology losing or gaining up to twenty minutes a day, it was practically useless on day two of the journey. All across Europe, bounties were offered for anyone who could solve the problem of determining longitude at sea: Philip III of Spain offered a life pension in ducats, while the famous Longitude Prize in England promised more than a million dollars in today's currency. The urgency of the problem – and the economic rewards for solving it – brought Galileo's mind back to the pursuit of 'equal time' that had first captured his imagination at the age of nineteen. His astronomical observations had suggested that the regular eclipses of Jupiter's moons might be useful for navigators keeping time at sea, but the method he devised was too complicated (and not as accurate as he had hoped). And so he returned, one last time, to the pendulum.

Fifty-eight years in the making, his slow hunch about the pendulum's 'magical property' had finally begun to take shape. The idea lay at the intersection point of multiple disciplines and interests: Galileo's memory of the altar lamp, his studies of motion and the moons of Jupiter, the rise of a global shipping industry, and its new demand for clocks that would be accurate to the second. Physics, astronomy, maritime navigation, and the daydreams of a college student: all these different strains converged in Galileo's mind. Aided by his son, he began drawing up plans for the first pendulum clock.

By the end of the next century, the pendulum clock had become a regular sight throughout Europe, particularly in England – in workplaces, town squares, even well-to-do homes. The British historian E. P. Thompson, in a brilliant essay on time and industrialization published in the late 1960s, noted that in the literature of the period, one of the telltale signs that a character has raised himself a rung or two up the socioeconomic ladder is the acquisition of a pocket watch. But these new timepieces were not just fashion accessories. A hundred times more accurate than its predecessors – losing or gaining only a minute or so a week – the pendulum clock brought about a change in the perception of time that we still live with today.

When we think about the technology that created the industrial revolution, we naturally conjure up the thunderous steam engines and steam-powered looms. But beneath the cacophony of the mills, a softer but equally important sound was everywhere: the ticking of pendulum clocks, quietly keeping time.

Imagine some alternative history where, for whatever reason, timekeeping technology lags behind the development of the other machines that catalyzed the industrial age. Would the industrial revolution have even happened? You can make a reasonably good case that the answer is no. Without clocks, the industrial takeoff that began in England in the middle of the eighteenth century would, at the very least, have taken much longer to reach escape velocity – for several reasons. Accurate clocks, thanks to their unrivaled ability to determine longitude at sea, greatly reduced the risks of global shipping networks, which gave the first industrialists a constant supply of raw materials and access to overseas markets. In the late 1600s and early 1700s, the most reliable watches in the world were manufactured in England, which created a pool of expertise with fine-tool manufacture that would prove to be incredibly handy when the demands of industrial innovation arrived, just as the glassmaking expertise producing spectacles opened the door for telescopes and microscopes. The watchmakers were the advance guard of what would become industrial engineering.

More than anything else, though, industrial life needed clock time to regulate the new working day. In older agrarian or feudal economies, units of time were likely to be described in terms of the time required to complete a task. The day was divided not into abstract, mathematical units, but into a series of activities: instead of fifteen minutes, time was described as how long it would take to milk the cow or nail soles to a new pair of shoes. Instead of being paid by the hour, craftsmen were conventionally paid by the piece produced – what was commonly called 'taken-work' – and their daily schedules were almost comically unregulated. Thompson cites the diary of one farming weaver from 1782 or 1783 as an example of scattered routines of pre-industrial work:

> On a rainy day, he might weave 8½ yards; on October 14th he carried his finished piece, and so wove only 4¾ yards; on the 23rd he worked out till 3 o'clock, wove two yards before the sun set . . . Apart from harvesting and threshing, churning, ditching and gardening, we have these entries: 'Wove 2½ yards the Cow having calved she required much attendance.' On January 25th he wove 2 yards, walked to a nearby village, and did 'sundry jobbs [*sic*] about the lathe and in the yard and wrote a letter in the evening.' Other occupations include jobbing with a horse and cart, picking cherries, working on a mill dam, attending a Baptist association, and a public hanging.

Try showing up for work in a modern office on that kind of clock. (Not even famously laid-back Google could tolerate that level of eccentricity.) For an industrialist trying to synchronize the actions of hundreds of workers with the mechanical tempo of the first factories, this kind of desultory work life was unmanageable. And so the creation of a viable industrial workforce required a profound reshaping of the human perception of time. The pottery manufacturer Josiah Wedgwood, whose Birmingham mills mark the very beginnings of industrial England, first implemented the convention of 'clocking in' to work each day. (The lovely double entendre of 'punching the clock' would have been meaningless to anyone born before 1700.) The whole idea of an 'hourly wage' – now practically universal in the modern world – came out of the time regimen of the industrial age. In such a system, Thompson writes, 'the employer must use the time of his labour, and see it is not wasted . . . Time is now currency: it is not passed but spent.'

For the first generations living through this transformation, the invention of 'time discipline' was deeply disorienting. Today, most of us in the developed world – and increasingly in the developing world – have been acclimated to the strict regimen of clock time from an early age. (Sit in on your average kindergarten classroom and you'll see the extensive focus on explaining and reinforcing the day's schedule.) The natural rhythms of tasks and leisure had to be forcibly replaced with an abstract grid. When you spend your whole life inside that grid, it seems like second nature, but when you are experiencing it for the first time, as the laborers of industrial England did in the second half of the eighteenth century, it arrives as a shock to the system. Timepieces were not just tools to help you coordinate the day's events, but something more ominous: the 'deadly statistical clock,' in Dickens's *Hard Times*, 'which measured every second with a beat like a rap upon a coffin lid.'

Naturally, that new regimen provoked a backlash. Not so much from the working classes – who began operating within the dictates of clock time by demanding overtime wages or shorter workdays – but rather from the aesthetes. To be a

Romantic at the turn of the nineteenth century was in part to break from the growing tyranny of clock time: to sleep late, ramble aimlessly through the city, refuse to live by the 'statistical clocks' that governed economic life. In *The Prelude*, Wordsworth announces his break from the 'keepers of our time':

> The guides, the wardens of our faculties
> And stewards of our labour, watchful men
> And skillful in the usury of time
> Sages, who in their prescience would control
> all accidents, and to the very road
> which they have fashioned would confine us down
> like engines . . .

The time discipline of the pendulum clock took the informal flow of experience and nailed it to a mathematical grid. If time is a river, the pendulum clock turned it into a canal of evenly spaced locks, engineered for the rhythms of industry. Once again, an increase in our ability to measure things turned out to be as important as our ability to make them.

That power to measure time was not distributed evenly through society: pocket watches remained luxury items until the middle of the nineteenth century, when a Massachusetts cobbler's son named Aaron Dennison borrowed the new process of manufacturing armaments using standardized, interchangeable parts and applied the same techniques to watchmaking. At the time, the production of advanced watches involved more than a hundred distinct jobs: one person would make individual flea-sized screws, by turning a piece of steel on a thread; another would inscribe watch cases; and so on. Dennison had a vision of machines mass-producing identical tiny screws that could then be put into any watch of the same model, and machines that would engrave cases at precision speed. His vision took him through a bankruptcy or two, and earned him the nickname 'the Lunatic of Boston' in the local press. But eventually, in the early 1860s, he hit on the idea of making a cheaper watch, without the conventional jeweled ornamentation that traditionally adorned pocket watches. It would be the first watch targeted at the mass market, not just the well-to-do.

Dennison's 'Wm. Ellery' watch – named after one of the signers of the Declaration of Independence, William Ellery – became a breakout hit, particularly with the soldiers of the Civil War. More than 160,000 watches were sold; even Abraham Lincoln owned and carried a 'Wm. Ellery' watch. Dennison turned a luxury item into a must-have commodity. In 1850, the average pocket watch cost $40; by 1878, a Dennison unjeweled watch cost just $3.50.

With watches spiking in popularity across the country, a Minnesota railroad agent named Richard Warren Sears stumbled across a box of unwanted watches from a local jeweler, and turned a tidy profit selling them to other station agents. Inspired by his success, he partnered with a Chicago businessman named Alvah Roebuck, and together they launched a mail-order publication showcasing a range of watch designs: the Sears, Roebuck catalog. Those fifteen pounds of mail-order catalogs currently weighing down your mailbox? They all started with the must-have gadget of the late nineteenth century: the consumer-grade pocket watch.

*

When Dennison first started thinking about democratizing time in America, in one key respect the clocks of the period remained woefully irregular. *Local* time – in cities and towns across the United States – was now accurate to the second, if you consulted a public clock in a place where time discipline was particularly crucial. But there were literally thousands of distinct local times. Clock time had been democratized, but it had not yet been standardized. Thanks to Dennison, watches were spreading quickly through the system, but they were all running at different times. In the United States, each town and village ran at its own independent pace – with clocks synced to the sun's position in the sky. As you moved west or east, even a few miles, the shifting relationship to the sun would produce a different time on a sundial. You could be standing in one city at 6:00 p.m., but just three towns over, the correct time would be 6:05. If you asked what time it was 150 years ago, you would have received at least twenty-three different answers in the state of Indiana, twenty-seven in Michigan, and thirty-eight in Wisconsin.

The strangest thing about this irregularity is the fact that no one noticed it. You couldn't talk directly to someone three towns over, and it took an hour or two to get there by unreliable roads at low speeds. So a few minutes of fuzziness in the respective clocks of each town didn't even register. But once people (and information) began to travel faster, the lack of standardization suddenly became a massive problem. Telegraphs and railroads exposed the hidden blurriness of nonstandardized clock time, just as, centuries before, the invention of the book had exposed the need for spectacles among the first generation of European readers.

Trains moving east or west – longitudinally – travel faster than the sun moves through the sky. So for every hour you traveled on a train, you needed to adjust your watch by four minutes. In addition, each railroad was running on its own clock, which meant that making a journey in the nineteenth century took some formidable number crunching. You'd leave New York at 8:00 a.m. New York time, catching the 8:05 on Columbia Railroad time, and arrive in Baltimore three hours later, at 10:54 Baltimore time, which was, technically speaking, 11:05 Columbia Rail time, where you would wait ten minutes and then catch the 11:01 B&O train to Wheeling, West Virgina, which was, technically speaking again, the 10:49 train if you were on Wheeling time, and 11:10 if your watch was still keeping New York time. And the funny thing is, all those different times were the right ones, at least measured by the sun's position in the sky. What made time easily measured by sundial made it infuriating by railroad.

The British had dealt with this problem by standardizing the entire country on Greenwich Mean Time in the late 1840s, synchronizing railroad clocks by telegraph. (To this day, clocks in every air traffic control center and cockpit around the world report Greenwich time; GMT is the single time zone of the sky.) But the United States was too sprawling to run off of one clock, particularly after the transcontinental railroad opened in 1869. With eight thousand towns across the country, each on its own clock, and over a hundred thousand miles of railroad track connecting them, the need for some kind of standardized system became overwhelming. For several decades, various proposals circulated for standardizing U. S. time, but nothing solidified. The logistical hurdles of coordinating schedules and clocks were immense, and somehow standardized time seemed to spark a strange feeling of resentment among ordinary citizens, as though it were an act against nature itself. A Cincinnati paper editorialized against standard time: 'It is simply preposterous . . . Let the people of Cincinnati stick to the truth as it is written by the sun, moon and stars.'

The United States remained temporally challenged until the early 1880s, when a railroad engineer named William F. Allen took on the cause. As the editor of a guide to railroad timetables, Allen knew firsthand how Byzantine the existing time system was. At a railroad convention in St Louis in 1883, Allen presented a map that proposed a shift from fifty distinct railroad times to the four time zones that are still in use, more than a century later: Eastern, Central, Mountain, and Pacific. Allen designed the map so that the divisions between time zones zigzagged slightly to correspond to the points where the major railroad lines connected, instead of having the divisions run straight down meridian lines.

Persuaded by Allen's plan, the railroad bosses gave him just nine months to make his idea a reality. He launched an energetic campaign of letter-writing and arm-twisting to convince observatories and city councils. It was an extraordinarily challenging campaign, but somehow Allen managed to pull it off. On November 18, 1883, the United States experienced one of the strangest days in the history of clock time, what became known as 'the day of two noons.' Eastern Standard Time, as Allen had defined it, ran exactly four minutes behind local New York time. On that November day, the Manhattan church bells rang out the old New York noon, and then four minutes later, a second noon was announced by a second ringing: the very first 12:00 p.m., EST. The second noon was broadcast out across the country via telegraph, allowing railroad lines and town squares all the way to the Pacific to synchronize their clocks.

The very next year, GMT was set as the international clock (based on Greenwich being located on the prime meridian), and the whole globe was divided into time zones. The world had begun to break free from the celestial rhythms of the solar system. Consulting the sun was no longer the most accurate way to tell the time. Instead, pulses of electricity traveling by telegraph wire from distant cities kept our clocks in sync.

One of the strange properties of the measurement of time is that it doesn't belong neatly to a single scientific discipline. In fact, each leap forward in our ability to measure time has involved a handoff from one discipline to another. The shift from sundials to pendulum clocks relied on a shift from astronomy to dynamics, the physics of motion. The next revolution in time would depend on electromechanics. With each revolution, though, the general pattern remained the same: scientists discover some natural phenomenon that displays the propensity for keeping 'equal time' that Galileo had observed in the altar lamps, and before long a wave of inventors and engineers begin using that new tempo to synchronize their devices. In the 1880s, Pierre and Jacques Curie first detected a curious property of certain crystals, including quartz, the very same material that had been so revolutionary for the glassmakers of Murano: under pressure, these crystals could be made to vibrate at a remarkably stable frequency. (This property came to be known as 'piezoelectricity.') The effect was even more pronounced when an alternating current was applied to the crystal.

The quartz crystal's remarkable ability to expand and contract in 'equal time' was first exploited by radio engineers in the 1920s, who used it to lock radio transmissions to consistent frequencies. In 1928, W. A. Marrison of Bell Labs built the first clock that kept time from the regular vibrations of a quartz crystal. Quartz clocks lost or gained only a thousandth of a second per day, and were far less vulnerable to atmospheric changes in temperature or humidity, not to mention movement, than pendulum clocks. Once again, the accuracy with which we measured time had increased by several orders of magnitude.

For the first few decades after Marrison's invention, quartz clocks became the de facto timekeeping devices for scientific or industrial use; standard U. S. time was kept by quartz clocks starting in the 1930s. But by the 1970s, the technology had gotten cheap enough for a mass market, with the emergence of the first quartz-based wristwatches. Today, just about every consumer appliance that has a clock on it – microwaves, alarm clocks, wristwatches, automobile clocks – all run on the equal time of quartz piezoelectricity. That transformation was predictable enough. Someone invents a better clock, and the first iterations are too expensive for consumer use. But eventually the price falls, and the new clock enters mainstream life. No surprise there. Once again, the surprise comes from somewhere else, from some other field that wouldn't initially seem to be all that dependent on time. New ways of measuring create new possibilities for making. With quartz time, that new possibility was computation.

A microprocessor is an extraordinary technological achievement on many levels, but few are as essential as this: computer chips are masters of time discipline. Think of the coordination needs of the industrial factory: thousands of short, repetitive tasks performed in proper sequence by hundreds of individuals. A microprocessor requires the same kind of time discipline, only the units being coordinated are bits of information instead of the hands and bodies of millworkers. (When Charles Babbage first invented a programmable computer in the middle of the Victorian Age, he called the CPU 'the mill' for a reason.) And instead of thousands of operations per minute, the microprocessor is executing billions of calculations per second, while shuffling information in and out of other microchips on the circuit board. Those operations are all coordinated by a master clock, now almost without exception made of quartz. (This is why tinkering with your computer to make it go faster than it was engineered to run is called 'overclocking.') A modern computer is the assemblage of many different technologies and modes of knowledge: the symbolic logic of programming languages, the electrical engineering of the circuit board, the visual language of interface design. But without the microsecond accuracy of a quartz clock, modern computers would be useless.

The accuracy of the quartz clock made its pendulum predecessors seem hopelessly erratic. But it had a similar effect on the ultimate timekeepers: the earth and the sun. Once we started measuring days with quartz clocks, we discovered that the length of the day was not as reliable as we had thought. Days shortened or lengthened in semi-chaotic ways thanks to the drag of the tides on the surface of the planet, wind blowing over mountain ranges, or the inner motion of the earth's molten core. If we really wanted to keep exact time, we couldn't rely on the earth's rotation. We needed a better timepiece. Quartz let us 'see' that the seemingly equal times of a solar day weren't nearly as equal as we had assumed. It was, in a way, the deathblow to the pre-Copernican universe. Not only was the earth not the center of the universe, but its rotation wasn't even consistent enough to define a day accurately. A block of vibrating sand could do the job much better.

Keeping proper time is ultimately all about finding – or making – things that oscillate in consistent rhythms: the sun rising in the sky, the moon waxing and waning, the altar lamp, the quartz crystal. The discovery of the atom in the early days of the twentieth century – led by scientists such as Niels Bohr and Werner Heisenberg – set in motion a series of spectacular and deadly innovations in energy and weaponry: nuclear power plants, hydrogen bombs. But the new science of the atom also revealed a less celebrated, but equally significant, discovery: the most consistent oscillator known to man. Studying the behavior of electrons orbiting within a cesium atom, Bohr noticed that they moved with an astonishing regularity. Untroubled by the chaotic drag of mountain ranges or tides, the electrons tapped out a rhythm that was several orders of magnitude more reliable than the earth's rotation.

The first atomic clocks were built in the mid-1950s, and immediately set a new standard of accuracy: we were now capable of measuring nanoseconds, a thousand times more accurate than the microseconds of quartz. That leap forward was what ultimately enabled the International Conference of Weights and Measures in 1967 to declare that it was time to reinvent time. In the new era, the master time for the planet would be measured in atomic seconds: 'the duration of 9,192,631,770 periods of the radiation corresponding to the transition between the two hyperfine levels of the ground state of the caesium 133 atom.' A day was no longer the time it took the earth to complete one rotation. A day became 86,400 atomic seconds, ticked off on 270 synchronized atomic clocks around the world.

The old timekeepers didn't die off completely, though. Modern atomic clocks actually tick off the seconds using a quartz mechanism, relying on the cesium atom and its electrons to correct any random aberrations in the quartz timekeeping. And the world's atomic clocks are reset every year based on the chaotic drift of the earth's orbit, adding or gaining a second so that the atomic and solar rhythms don't get too far out of sync. The multiple scientific fields of time discipline – astronomy, electromechanics, subatomic physics – are all embedded within the master clock.

The rise of the nanosecond might seem like an arcane shift, interesting only to the sort of person who attends a conference on weights and measures. And yet everyday life has been radically transformed by the rise of atomic time.

Global air travel, telephone networks, financial markets – all rely on the nanosecond accuracy of the atomic clock. (Rid the world of these modern clocks, and the much vilified practice of high-frequency trading would disappear in a nanosecond.) Every time you glance down at your smartphone to check your location, you are unwittingly consulting a network of twenty-four atomic clocks housed in satellites in low-earth orbit above you. Those satellites are sending out the most elemental of signals, again and again, in perpetuity: the time is 11:48:25.084738 . . . the time is 11:48:25.084739 . . . When your phone tries to figure out its location, it pulls down at least three of these time stamps from satellites, each reporting a slightly different time thanks to the duration it takes the signal to travel from satellite to the GPS receiver in your hand. A satellite reporting a later time is closer than one reporting an earlier time. Since the satellites have perfectly predictable locations, the phone can calculate its exact position by triangulating among the three different time stamps. Like the naval navigators of the eighteenth century, GPS determines your location by comparing clocks. This is in fact one of the recurring stories of the history of the clock: each new advance in timekeeping enables a corresponding advance in our mastery of geography – from ships, to railroads, to air traffic, to GPS. It's an idea that Einstein would have appreciated: measuring time turns out to be key to measuring space.

The next time you glance down at your phone to check what time it is or where you are, the way you might have glanced at a watch or a map just two decades ago, think about the immense, layered network of human ingenuity that has been put in place to make that gesture possible. Embedded in your ability to tell the time is the understanding of how electrons circulate within cesium atoms; the knowledge of how to send microwave signals from satellites and how to measure the exact speed with which they travel; the ability to position satellites in reliable orbits above the earth, and of course the actual rocket science needed to get them off the ground; the ability to trigger steady vibrations in a block of silicon dioxide – not to mention all the advances in computation and microelectronics and network science necessary to process and represent that information on your phone. You don't need to know any of these things to tell the time now, but that's the way progress works: the more we build up these vast repositories of scientific and technological understanding, the more we conceal them. Your mind is silently assisted by all that knowledge each time you check your phone to see what time it is, but the knowledge itself is hidden from view. That is a great convenience, of course, but it can obscure just how far we've come since Galileo's altar-lamp daydreams in the Duomo di Pisa.

At first glance, the story of time's measurement would seem to be all about acceleration, dividing up the day into smaller and smaller increments so that we can move things faster: bodies, dollars, bits. But time in the atomic age has also moved in the exact opposite direction: slowing things down, not speeding them up; measuring in eons, not microseconds. In the 1890s, while working on her doctoral thesis in Paris, Marie Curie proposed for the first time that radiation was not some kind of chemical reaction between molecules, but something intrinsic to the atom – a discovery so critical to the development of physics, in fact, that she would become the first woman ever to win a Nobel Prize. Her research quickly drew the attention of her husband, Pierre Curie, who abandoned his own research into crystals to focus on radiation. Together they discovered that radioactive elements decayed at constant rates. The half-life of carbon 14, for instance, is 5,730 years. Leave some carbon 14 lying around for five thousand years or so, and you'll find that half of it is gone.

Once again, science had discovered a new source of 'equal time' – only this clock wasn't ticking out the microseconds of quartz oscillations, or the nanoseconds of cesium electrons. Radiocarbon decay was ticking on the scale of centuries or millennia. Pierre Curie had surmised that the decay rate of certain elements might be used as a 'clock' to determine the age of rocks. But the technique, now popularly known as carbon dating, wasn't perfected until the late 1940s. Most clocks focus on measuring the present: What time is it right now? But radiocarbon clocks are all about the past. Different elements decay at wildly different rates, which means that they are like clocks running at different time scales. Carbon 14 'ticks' every five thousand years, but potassium 40 'ticks' every 1.3 billion years. That makes radiocarbon dating an ideal clock for the deep time of human history, while potassium 40 measures geologic time, the history of the planet itself. Radiometric dating has been critical in determining the age of the earth itself, establishing the most convincing scientific evidence that the biblical story of the earth being six thousand years old is just that: a story, not fact. We have immense knowledge about the prehistoric migrations of humans across the globe in large part thanks to carbon dating. In a sense, the 'equal time' of radioactive decay has turned prehistoric time into history. When *Homo sapiens* first crossed the Bering Land Bridge into the Americas more than ten thousand years ago, there were no historians capable of writing down a narrative account of their journey. Yet their story was nonetheless captured by the carbon in their bones and the charcoal deposits they left behind at campsites. It was a story written in the language of atomic physics. But we couldn't read that story without a new kind of clock. Without radiometric dating, 'the deep time' of human migrations or geologic change would be like a history book where all the pages have been randomly shuffled: teeming with facts but lacking chronology and causation. Knowing what time it was turned that raw data into meaning.

High in the Southern Snake Mountains in eastern Nevada, a grove of bristlecone pines grows in the dry, alkaline soil. The pines are small trees for conifers, rarely more than thirty feet high, gnarled by the constant winds rolling across the desert range. We know from carbon dating (and tree rings) that some of them are more than five thousand years old – the oldest living things on the planet.

At some point, several years from now, a clock will be buried in the soil beneath those pines, a clock designed to measure time on the scale of civilizations, not seconds. It will be – as its primary designer, the computer scientist Danny Hillis, puts it – 'a clock that ticks once a year. The century hand advances once every 100 years, and the cuckoo comes out on the millennium.' It is being engineered to keep time for at least ten thousand years, roughly the length of human civilization to date. It is an exercise in a different kind of time discipline: the discipline of avoiding short-term thinking, of forcing yourself to think about our actions and their consequences on the scale of centuries and millennia. Borrowing a wonderful phrase from the musician and artist Brian Eno, the device is called 'the Clock of the Long Now.'

The organization behind this device, the Long Now Foundation – cofounded by Hillis, Eno, Stewart Brand, and a few other visionaries – aims to build a number of ten-thousand-year clocks. (The first one is being constructed for a mountainside location in West Texas.) Why go to such extravagant lengths to build a clock that might tick only once in your lifetime? Because new modes of measuring force us to think about the world in a new light. Just as the microseconds of quartz and cesium opened up new ideas that transformed everyday life in countless ways, the slow time of the Long Now clock helps us think in new ways about the future. As Long Now board member Kevin Kelly puts it:

> If you have a Clock ticking for 10,000 years what kinds of generational-scale questions and projects will it suggest? If a Clock can keep going for ten millennia, shouldn't we make sure our civilization does as well? If the Clock keeps going after we are personally long dead, why not attempt other projects that require future generations to finish? The larger question is, as virologist Jonas Salk once asked, 'Are we being good ancestors?'

This is the strange paradox of time in the atomic age: we live in ever shorter increments, guided by clocks that tick invisibly with immaculate precision; we have short attention spans and have surrendered our natural rhythms to the abstract grid of clock time. And yet simultaneously, we have the capacity to imagine and record histories that are thousands or millions of years old, to trace chains of cause and effect that span dozens of generations. We can wonder what time it is and glance down at our phone and get an answer that is accurate to the split-second, but we can also appreciate that the answer was, in a sense, five hundred years in the making: from Galileo's altar lamp to Niels Bohr's cesium, from the chronometer to *Sputnik*. Compared to an ordinary human being from Galileo's age, our time horizons have expanded in both directions: from the microsecond to the millennium.

Which measure of time will win out in the end: our narrow focus on the short term, or our gift for the long now? Will we be high-frequency traders or good ancestors? For that question, only time will tell.

CHAPTER 6

Light

Imagine some alien civilization viewing Earth from across the galaxies, looking for signs of intelligent life. For millions of years, there would be almost nothing to report: the daily flux of weather moving across the planet, the creep of glaciers spreading and retreating every hundred thousand years or so, the incremental drift of continents. But starting about a century ago, a momentous change would suddenly be visible: at night, the planet's surface would glow with the streetlights of cities, first in the United States and Europe, then spreading steadily across the globe, growing in intensity. Viewed from space, the emergence of artificial lighting would arguably have been the single most significant change in the planet's history since the Chicxulub asteroid collided with Earth sixty-five million years ago, coating the planet in a cloud of superheated ash and dust. From space, all the transformations that marked the rise of human civilization would be an afterthought: opposable thumbs, written language, the printing press – all these would pale beside the brilliance of *Homo lumens*.

Viewed from the surface of the earth, of course, the

invention of artificial light had more rivals in terms of visible innovations, but its arrival marked a threshold point in human society. Today's night sky now shines six thousand times brighter than it did just 150 years ago. Artificial light has transformed the way we work and sleep, helped create global networks of communication, and may soon enable radical breakthroughs in energy production. The lightbulb is so bound up in the popular sense of innovation that it has become a metaphor for new ideas themselves: the 'lightbulb' moment has replaced Archimedes's eureka as the expression most likely to be invoked to celebrate a sudden conceptual leap.

One of the odd things about artificial light is how stagnant it was as a technology for centuries. This is particularly striking given that artificial light arrived via the very first technology, when humans originally mastered the controlled fire more than a hundred thousand years ago. The Babylonians and Romans developed oil-based lamps, but that technology virtually disappeared during the (appropriately named) Dark Ages. For almost two thousand years, all the way to the dawn of the industrial age, the candle was the reigning solution for indoor lighting. Candles made from beeswax were highly prized but too expensive for anyone but the clergy or the aristocracy. Most people made do with tallow candles, which burned animal fat to produce a tolerable flicker, accompanied by a foul odor and thick smoke.

As our nursery rhymes remind us, candle-making was a popular vocation during this period. Parisian tax rolls from 1292 listed seventy-two 'chandlers,' as they were called, doing business in the city. But most ordinary households made their own tallow candles, an arduous process that could go on for days: heating up containers of animal fat, and dipping wicks into them. In a diary entry from 1743, the president of Harvard noted that he had produced seventy-eight pounds of tallow candles in two days of work, a quantity that he managed to burn through two months later.

It's not hard to imagine why people were willing to spend so much time manufacturing candles at home. Consider what life would have been like for a farmer in New England in 1700. In the winter months the sun goes down at five, followed by fifteen hours of darkness before it gets light again. And when that sun goes down, it's pitch-black: there are no streetlights, flashlights, lightbulbs, fluorescents – even kerosene lamps haven't been invented yet. There's just a flickering glow of a fireplace, and the smoky burn of the tallow candle.

Those nights were so oppressive that scientists now believe our sleep patterns were radically different in the ages before ubiquitous night lighting. In 2001, the historian Roger Ekirch published a remarkable study that drew upon hundreds of diaries and instructional manuals to convincingly argue that humans had historically divided their long nights into two distinct sleep periods. When darkness fell, they would drift into 'first sleep,' waking after four hours to snack, relieve themselves, have sex, or chat by the fire, before heading back for another four hours of 'second sleep.' The lighting of the nineteenth century disrupted this ancient rhythm, by opening up a whole array of modern activities that could be performed after sunset: everything from theaters and restaurants to factory labor. Ekirch documents the way the ideal of a single, eight-hour block of continuous sleep was constructed by nineteenth-century customs, an adaptation to a dramatic change in the lighting environment of human settlements. Like all adaptations, its benefits carried inevitable costs: the middle-of-the-night insomnia that plagues millions of people around the world is not, technically speaking, a disorder, but rather the body's natural sleep rhythms asserting themselves over the prescriptions of nineteenth-century convention. Those waking moments at three a.m. are a kind of jet lag caused by artificial light instead of air travel.

The flicker of the tallow candle had not been strong enough to transform our sleep patterns. To make a cultural change that momentous, you needed the steady bright glow of nineteenth-century lighting. By the end of the century, that light would come from the burning filaments of the electric lightbulb. But the first great advance in the century of light came from a source that seems macabre to us today: the skull of a fifty-ton marine mammal.

It's a story that begins with a storm. Legend has it that sometime around 1712 a powerful nor'easter off the coast of Nantucket blew a ship captain named Hussey far out to sea. In the deep waters of the North Atlantic, he encountered one of Mother Nature's most bizarre and intimidating creations: the sperm whale.

Hussey managed to harpoon the beast – though some skeptics think it simply washed ashore in the storm. Either way, when locals dissected the giant mammal, they discovered something utterly bizarre: inside the creature's massive head, they found a cavity above the brain, filled with a white, oily substance. Thanks to its resemblance to seminal fluid, the whale oil came to be called 'spermaceti.'

To this day, scientists are not entirely sure why sperm whales produce spermaceti in such vast quantities. (A mature sperm whale holds as much as five hundred gallons inside its skull.) Some believe the whales use the spermaceti for buoyancy; others believe it helps with the mammal's echolocation system. New Englanders, however, quickly discovered another use for spermaceti: candles made from the substance produce a much stronger, whiter light than tallow candles, without the offensive smoke. By the second half of the eighteenth century, spermaceti candles had become the most prized form of artificial light in America and Europe.

In a 1751 letter, Ben Franklin described how much he enjoyed the way the new candles 'afford a clear white Light; may be held in the Hand, even in hot Weather, without softening; that their Drops do not make Grease Spots like those from common Candles; that they last much longer, and need little or no Snuffing.' Spermaceti light quickly became an expensive habit for the well-to-do. George Washington estimated that he spent $15,000 a year in today's currency burning spermaceti candles. The candle business became so lucrative that a group of manufacturers formed an organization called United Company of Spermaceti Chandlers, conventionally known as the 'Spermaceti Trust,' designed to keep competitors out of the business and force the whalers to keep their prices in check.

Despite the candle-making monopoly, significant economic rewards awaited anyone who managed to harpoon a sperm whale. The artificial light of the spermaceti candle triggered an explosion in the whaling industry, building out the beautiful seaside towns of Nantucket and Edgartown. But as elegant as these streets seem today, whaling was a dangerous and repulsive business. Thousands of lives were lost at sea chasing these majestic creatures, including from the notorious sinking of the *Essex*, which ultimately inspired Herman Melville's masterpiece, *Moby-Dick*.

Extracting the spermaceti was almost as difficult as harpooning the whale itself. A hole would be carved in the side of the whale's head, and men would crawl into the cavity above the brain – spending days inside the rotting carcass, scraping spermaceti out of the brain of the beast. It's remarkable to think that only two hundred years ago, this was the reality of artificial light: if your great-great-great-grandfather wanted to read his book after dark, some poor soul had to crawl around in a whale's head for an afternoon.

Somewhere on the order of three hundred thousand sperm whales were slaughtered in just a little more than a century. It is likely that the entire population would have been killed off had we not found a new source of oil for artificial light in the ground, introducing petroleum-based solutions such as the kerosene lamp and the gaslight. This is one of the stranger twists in the history of extinction: because humans discovered deposits of ancient plants buried deep below the surface of the earth, one of the ocean's most extraordinary creatures was spared.

Fossil fuels would become central to almost all aspects of twentieth-century life, but their first commercial use revolved around light. These new lamps were twenty times brighter than any candle had ever been, and their superior brightness helped spark an explosion in magazine and newspaper publishing in the second half of the nineteenth century, as the dark hours after work became increasingly compatible with reading. But they also sparked literal explosions: thousands of people died each year in the fiery eruption of a reading light.

Despite these advances, artificial light remained extremely expensive by modern standards. In today's society, light is comparatively cheap and abundant; 150 years ago, reading after dark was a luxury. The steady march of artificial light since then, from a rare and feeble technology to a ubiquitous and powerful one, gives us one map for the path of progress over that period. In the late 1990s, the Yale historian William D. Nordhaus published an ingenious study that charted that path in extraordinary detail, analyzing the true costs of artificial light over thousands of years of innovation.

When economic historians try to gauge the overall health of economies over time, average wages are usually one of the first places they start. Are people today making more money than they did in 1850? Of course, inflation makes those comparisons tricky: someone making ten dollars a day was upper-middle-class in nineteenth-century dollars. That's why we have inflation tables that help us understand that ten dollars then is worth $160 in today's currency. But inflation covers only part of the story. 'During periods of major technological change,' Nordhaus argued, 'the construction of accurate price indexes that capture the impact of new technologies on living standards is beyond the practical capability of official statistical agencies. The essential difficulty arises for the obvious but usually overlooked reason that most of the goods we consume today were not produced a century ago.' Even if you had $160 in 1850, you couldn't buy a wax phonograph, not to mention an iPod. Economists and historians needed to factor not only the general value of a currency, but also some sense of what that currency could buy.

This is where Nordhaus proposed using the history of artificial light to illuminate the true purchasing power of wages over the centuries. The vehicles of artificial light vary dramatically over the years: from candles to LEDs. But the light they produce is a constant, a kind of anchor in the storm of rapid technological innovation. So Nordhaus proposed as his unit of measure the cost of producing one thousand 'lumen-hours' of artificial light.

A tallow candle in 1800 would cost roughly forty cents per thousand lumen-hours. A fluorescent bulb in 1992, when Nordhaus originally compiled his research, cost a tenth of a cent for the same amount of light. That's a four-hundred-fold increase in efficiency. But the story is even more dramatic when you compare those costs to the average wages from the period. If you worked for an hour at the average wage of 1800, you could buy yourself ten minutes of artificial light. With kerosene in 1880, the same hour of work would give you three hours of reading at night. Today, you can buy three hundred *days* of artificial light with an hour of wages.

Something extraordinary obviously happened between the days of tallow candles or kerosene lamps and today's illuminated wonderland. That something was the electric lightbulb.

The strange thing about the electric lightbulb is that it has come to be synonymous with the 'genius' theory of innovation – the single inventor inventing a single thing, in a moment of sudden inspiration – while the true story behind its creation actually makes the case for a completely different explanatory framework: the network/systems model of innovation. Yes, the lightbulb marked a threshold in the history of innovation, but for entirely different reasons. It would be pushing things to claim that the lightbulb was crowdsourced, but it is even more of a distortion to claim that a single man named Thomas Edison invented it.

The canonical story goes something like this: after a triumphant start to his career inventing the phonograph and the stock ticker, a thirty-one-year-old Edison takes a few months off to tour the American West – perhaps not coincidentally, a region that was significantly darker at night than the gaslit streets of New York and New Jersey. Two days after returning to his lab in Menlo Park, in August 1878, he draws three diagrams in his notebook and titles them 'Electric Light.' By 1879, he files a patent application for an 'electric lamp' that displays all the main characteristics of the lightbulb we know today. By the end of 1882, Edison's company is powering electric light for the entire Pearl Street district in Lower Manhattan.

It's a thrilling story of invention: the young wizard of Menlo Park has a flash of inspiration, and within a few years his idea is lighting up the world. The problem with this story is that people had been inventing incandescent light for eighty years before Edison turned his mind to it. A lightbulb involves three fundamental elements: some kind of filament that glows when an electric current runs through it, some mechanism to keep the filament from burning out too quickly, and a means of supplying electric power to start the reaction in the first place. In 1802, the British chemist Humphry Davy had attached a platinum filament to an early electric battery, causing it to burn brightly for a few minutes. By the 1840s, dozens of separate inventors were working on variations of the lightbulb. The first patent was issued in 1841 to an Englishman named Frederick de Moleyns. The historian Arthur A. Bright compiled a list of the lightbulb's partial inventors, leading up to Edison's ultimate triumph in the late 1870s.

Date	Inventor	Nationality	Element	Atmosphere
1838	*Jobard*	Belgian	Carbon	Vacuum
1840	*Grove*	English	Platinum	Air
1841	*De Moleyns*	English	Carbon	Vacuum
1845	*Starr*	American	Platinum carbon	Air Vacuum
1848	*Staite*	English	Platinum/ iridium	Air
1849	*Petrie*	American	Carbon	Vacuum
1850	*Shepard*	American	Iridium	Air
1852	*Roberts*	English	Carbon	Vacuum
1856	*De Changy*	French	Platinum Carbon	Air Vacuum
1858	*Gardiner & Blossom*	American	Platinum	Vacuum

Date	Inventor	Nationality	Element	Atmosphere
1859	*Farmer*	American	Platinum	Air
1860	*Swan*	English	Carbon	Vacuum
1865	*Adams*	American	Carbon	Vacuum
1872	*Lodyguine*	Russian	Carbon	Vacuum
			Carbon	Nitrogen
1875	*Kosloff*	Russian	Carbon	Nitrogen
1876	*Bouliguine*	Russian	Carbon	Vacuum
1878	*Fontaine*	French	Carbon	Vacuum
1878	*Lane-Fox*	English	Platinum/ iridium	Nitrogen
			Platinum/ iridium	Air
			Asbestos/ carbon	Nitrogen
1878	*Sawyer*	American	Carbon	Nitrogen
1878	*Maxim*	American	Carbon	Hydrocarbon
1878	*Farmer*	American	Carbon	Nitrogen
1879	*Farmer*	American	Carbon	Vacuum
1879	*Swan*	English	Carbon	Vacuum
1879	*Edison*	American	Carbon	Vacuum

At least half of the men had hit upon the basic formula that Edison ultimately arrived at: a carbon filament, suspended in a vacuum to prevent oxidation, thus keeping the filament from burning up too quickly. In fact, when Edison finally began tinkering with electric light, he spent months working on a feedback system for regulating the flow of electricity to prevent melting, before finally abandoning that approach in favor of the vacuum – despite the fact that nearly half of his predecessors had already embraced the vacuum as the best environment for a sustained glow. The lightbulb was the kind of innovation that comes together over decades, in pieces. There was no lightbulb moment in the story of the lightbulb. By the time Edison flipped the switch at the Pearl Street station, a handful of other firms were already selling their own models of incandescent electric lamps. The British inventor Joseph Swan had begun lighting homes and theaters a year earlier. Edison invented the lightbulb the way Steve Jobs invented the MP3 player: he wasn't the first, but he *was* the first to make something that took off in the marketplace.

So why does Edison get all the credit? It's tempting to use the same backhanded compliment that many leveled against Steve Jobs: that he was a master of marketing and PR. It is true that Edison had a very tight relationship with the press at this point of his career. (On at least one occasion, he gave shares in his company to a journalist in exchange for better coverage.) Edison was also a master of what we would now call 'vaporware': He announced nonexistent products to scare off competitors. Just a few months after he had started work on electric light, he began telling reporters from New York papers that the problem had been solved, and that he was on the verge of launching a national system of magical electrical light. A system so simple, he says, 'that a bootblack might understand it.'

Despite this bravado, the fact remained that the finest specimen of electric light in the Edison lab couldn't last five minutes. But that didn't stop him from inviting the press out to Menlo Park lab to see his revolutionary lightbulb. Edison would bring each reporter in one at a time, flick the switch on a bulb, and let the reporter enjoy the light for three or four minutes before ushering him from the room. When he asked how long his lightbulbs would last, he answered confidently: 'Forever, almost.'

But for all this bluffing, Edison and his team did manage to ship a revolutionary and magical product, as the Apple marketing might have called the Edison lightbulb. Publicity and marketing will only get you so far. By 1882, Edison had produced a lightbulb that decisively outperformed its competitors, just as the iPod outperformed its MP3-player rivals in its early years.

In part, Edison's 'invention' of the lightbulb was less about a single big idea and more about sweating the details. (His famous quip about invention being one percent inspiration and ninety-nine percent perspiration certainly holds true for his adventures in artificial light.) Edison's single most significant contribution to the electric lightbulb itself was arguably the carbonized bamboo filament he eventually settled on. Edison wasted at least a year trying to make platinum work as a filament, but it was too expensive and prone to melting. Once he abandoned platinum, Edison and his team tore through a veritable botanic garden of different materials: 'celluloid, wood shavings (from boxwood, spruce, hickory, baywood, cedar, rosewood, and maple), punk, cork, flax, coconut hair and shell, and a variety of papers.' After a year of experimentation, bamboo emerged as the most durable substance, which set off one of the strangest chapters in the history of global commerce. Edison dispatched a series of Menlo Park emissaries to scour the globe for the most incandescent bamboo in the natural world. One representative paddled down two thousand miles of river in Brazil.

Another headed to Cuba, where he was promptly struck down with yellow fever and died. A third representative named William Moore ventured to China and Japan, where he struck a deal with a local farmer for the strongest bamboo the Menlo Park wizards had encountered. The arrangement remained intact for many years, supplying the filaments that would illuminate rooms all over the world. Edison may not have invented the lightbulb, but he did inaugurate a tradition that would turn out to be vital to modern innovation: American electronics companies importing their component parts from Asia. The only difference is that, in Edison's time, the Asian factory was a forest.

The other key ingredient to Edison's success lay in the team he had assembled around him in Menlo Park, memorably known as the 'muckers.' The muckers were strikingly diverse both in terms of professional expertise and nationality: the British mechanic Charles Batchelor, the Swiss machinist John Kruesi, the physicist and mathematician Francis Upton, and a dozen or so other draftsmen, chemists, and metalworkers. Because the Edison lightbulb was not so much a single invention as a bricolage of small improvements, the diversity of the team turned out to be an essential advantage for Edison. Solving the problem of the filament, for instance, required a scientific understanding of electrical resistance and oxidation that Upton provided, complementing Edison's more untutored, intuitive style; and it was Batchelor's mechanical improvisations that enabled them to test so many different candidates for the filament itself. Menlo Park marked the beginning of an organizational form that would come to prominence in the twentieth century: the cross-disciplinary research-and-development lab. In this sense, the transformative ideas and technologies that came out of places such as Bell Labs and Xerox-PARC have their roots in Edison's workshop. Edison didn't just invent technology; he invented an entire system for inventing, a system that would come to dominate twentieth-century industry.

Edison also helped inaugurate another tradition that would become vital to contemporary high-tech innovation: paying his employees in equity rather than just cash. In 1879, in the middle of the most frenetic research into the lightbulb, Edison offered Francis Upton stock worth 5 percent of the Edison Electric Light Company – though Upton would have to forswear his salary of $600 a year. Upton struggled over the decision, but ultimately decided to take the stock grant, over the objections of his more fiscally conservative father. By the end of the year, the ballooning value of Edison stock meant that his holdings were already worth $10,000, more than a million dollars in today's currency. Not entirely graciously, Upton wrote to his father, 'I cannot help laughing when I think how timid you were at home.'

By any measure, Edison was a true genius, a towering figure in nineteenth-century innovation. But as the story of the lightbulb makes clear, we have historically misunderstood that genius. His greatest achievement may have been the way he figured out how to make *teams* creative: assembling diverse skills in a work environment that valued experimentation and accepted failure, incentivizing the group with financial rewards that were aligned with the overall success of the organization, and building on ideas that originated elsewhere. 'I am not overly impressed by the great names and reputations of those who might be trying to beat me to an invention . . . It's their "ideas" that appeal to me,' Edison famously said. 'I am quite correctly described as "more of a sponge than an inventor."'

The lightbulb was the product of networked innovation, and so it is probably fitting that the reality of electric light ultimately turned out to be more of a network or system than a single entity. The true victory lap for Edison didn't come with that bamboo filament glowing in a vacuum; it came with the lighting of the Pearl Street district two years later. To make that happen, you needed to invent lightbulbs, yes, but you also needed a reliable source of electric current, a system for distributing that current through a neighborhood, a mechanism for connecting individual lightbulbs to the grid, and a meter to gauge how much electricity each household was using. A lightbulb on its own is a curiosity piece, something to dazzle reporters with. What Edison and the muckers created was much bigger than that: a network of multiple innovations, all linked together to make the magic of electric light safe and affordable.

Why should we care whether Edison invented the lightbulb as a lone genius or as part of a wider network? For starters, if the invention of the lightbulb is going to be a canonical story of how new technologies come into being, we might as well tell an accurate story. But it's more than just a matter of getting the facts right, because there are social and political implications to these kinds of stories. We know that one key driver of progress and standards of living is technological innovation. We know that we want to encourage the trends that took us from ten minutes of artificial light on one hour's wage to three hundred days. If we think that innovation comes from a lone genius inventing a new technology from scratch, that model naturally steers us toward certain policy decisions, like stronger patent protection. But if we think that innovation comes out of collaborative networks, then we want to support different policies and organizational forms: less rigid patent laws, open standards, employee participation in stock plans, cross-disciplinary connections. The lightbulb shines light on more than just our bedside reading; it helps us see more clearly the way new ideas come into being, and how to cultivate them as a socicty.

Artificial light turns out to have an even deeper connection to political values. Just six years after Edison lit the Pearl Street district, another maverick would push the envelope of light in a new direction, while walking the streets just a few blocks north of Edison's illuminated wonderland. The muckers might have invented the system of electric light, but the next breakthrough in artificial light would come from a muckraker.

Buried deep near the center of the Great Pyramid of Giza lies a granite-faced cavity known as 'the King's Chamber.' The room contains only one object: an open rectangular box, sometimes called a 'coffer,' carved out of red Aswan granite, chipped on one corner. The chamber's name derives from the assumption that the coffer had been a sarcophagus that once contained the body of Khufu, the pharaoh who built the pyramid more than four thousand years ago. But a long line of maverick Egyptologists have suggested that the coffer had other uses. One still-circulating theory notes that the coffer possesses the exact dimensions that the Bible attributes to the original Ark of the Covenant, suggesting to some that the coffer once housed the legendary Ark itself.

In the fall of 1861, a visitor came to the King's Chamber in the throes of an equally outlandish theory, this one revolving around a different Old Testament ark. The visitor was Charles Piazzi Smyth, who for the preceding fifteen years had served as the Royal Astronomer of Scotland, though he was a classic Victorian polymath with dozens of eclectic interests. Smyth had recently read a bizarre tome that contended that the pyramids had been originally built by the biblical Noah. Long an armchair Egyptologist, Smyth had grown so obsessed with the theory that he left his armchair in Edinburgh and headed off to Giza to do his own investigations firsthand. His detective work would ultimately lead to a bizarre stew of numerology and ancient history, published in a series of books and pamphlets over the coming years. Smyth's detailed analyses of the pyramid's structure convinced him that the builders had relied on a unit of measurement that was almost exactly equivalent to the modern British inch. Smyth interpreted this correspondence to be a sign that the inch itself was a holy measure, passed directly from God to Noah himself. This in turn gave Smyth the artillery he needed to attack the metric system that had begun creeping across the English Channel. The revelation of the Egyptian inch made it clear that the metric system was not just a symptom of malevolent French influence. It was also a betrayal of divine will.

Smyth's scientific discoveries in the Great Pyramid may

not have stood the test of time, or even kept Britain from going metric. Yet he still managed to make history in the King's Chamber. Smyth brought the bulky and fragile tools of wet-plate photography (then state of the art) to Giza to document his findings. But the collodion-treated glass plates couldn't capture a legible image in the King's Chamber, even when the room was lit by torchlight. Photographers had tinkered with artificial lighting since the first daguerreotypes were printed in the 1830s, but almost all the solutions to date had produced unsatisfactory results. (Candles and gaslight were useless, obviously.) Early experiments heated a ball of calcium carbonate – the 'limelight' that would illuminate theater productions until the dawn of electric light – but limelit photographs suffered from harsh contrasts and ghostly white faces.

The failed experiments with artificial lighting meant that by the time Smyth set up his gear in the King's Chamber, more than thirty years after the invention of the daguerreotype, the art of photography was still entirely dependent on natural sunlight, a resource that was not exactly abundant in the inner core of a massive pyramid. But Smyth had heard of recent experiments using wire made of magnesium – photographers who twisted the wire into a bow and set it ablaze before capturing their low-light image. The technique was promising, but the light was unstable and generated an unpleasant amount of dense fumes. Burning magnesium wire in a closed environment had a tendency to make ordinary portraits look as though they were composed in dense fog.

Smyth realized that what he needed in the King's Chamber was something closer to a flash than a slow burn. And so – for the first time in history, as far as we know – he mixed magnesium with ordinary gunpowder, creating a controlled mini-explosion that illuminated the walls of the King's Chamber for a split second, allowing him to record its secrets on his glass plates. Today, the tourists that pass through the Great Pyramid encounter signs that forbid the use of flash photography inside the vast structure. They do not mention that the Great Pyramid also marks the site where flash photography was invented.

Or at least, *one* of the sites where flash photography was invented. Just as with Edison's lightbulb, the true story of flash photography's origins is a more complicated, more networked affair. Big ideas coalesce out of smaller, incremental breakthroughs. Smyth may have been the first to conceive of the idea of combining magnesium with an oxygen-rich, combustible element, but flash photography itself didn't become a mainstream practice for another two decades, when two German scientists, Adolf Miethe and Johannes Gaedicke, mixed fine magnesium powder with potassium chlorate, creating a much more stable concoction that allowed high-shutter-speed photographs in low-light conditions. They called it *Blitzlicht* – literally, 'flash light.'

Word of Miethe and Gaedicke's invention soon trickled out of Germany. In October 1887, a New York paper ran a four-line dispatch about *Blitzlicht*. It was hardly a front-page story; the vast majority of New Yorkers ignored it altogether. But the idea of flash photography set off a chain of associations in the mind of one reader – a police reporter and amateur photographer who stumbled across the article while having breakfast with his wife in Brooklyn. His name was Jacob Riis.

Then a twenty-eight-year-old Danish immigrant, Riis would ultimately enter the history books as one of the original muckrakers of the late nineteenth century, the man who did more to expose the squalor of tenement life – and inspire a progressive reform movement – than any other figure of the era. But until that breakfast in 1887, Riis's attempts to shine light on the appalling conditions in the slums of Manhattan had failed to change public opinion in any meaningful way. A close confidant of then police commissioner Teddy Roosevelt, Riis had been exploring the depths of Five Points and other Manhattan hovels for years. With over half a million people living in only fifteen thousand tenements, sections of Manhattan were the most densely populated places on the planet. Riis was fond of taking late-night walks through the bleak alleyways on his way back home to Brooklyn from the police headquarters on Mulberry Street. 'We used to go in the small hours of the morning,' he later recalled, 'into the worst tenements to count noses and see if the law against overcrowding was violated, and the sights I saw there gripped my heart until I felt that I must tell of them, or burst, or turn anarchist, or something.'

Appalled by what he had discovered on his expeditions, Riis began writing about the mass tragedy of the tenements for local papers and national magazines such as *Scribner's* and *Harper's Weekly*. His written accounts of the shame of the cities belonged to a long tradition, dating back at least to Dickens's horrified visit to New York in 1840. A number of exhaustive surveys of tenement depravity had been published over the years, with titles like 'The Report of the Council of Hygiene and Public Health.' An entire genre of 'sunshine and shadow' guidebooks to Five Points and its ilk flourished after the Civil War, offering curious visitors tips on exploring the seedy underbelly of big-city life, or at least exploring it vicariously from the safety of a small-town oasis. (The phrase 'slumming it' originates with these tourist expeditions.) But despite stylistic differences, these texts shared one attribute: they had almost no effect on improving the actual living conditions of those slum dwellers.

Riis had long suspected that the problem with tenement reform – and urban poverty initiatives generally – was ultimately a problem of imagination. Unless you walked through the streets of Five Points after midnight, or descended into the dark recesses of interior apartments populated by multiple families at a time, you simply couldn't imagine the conditions; they were too far removed from the day-to-day experience of most Americans, or at least most voting Americans. And so the political mandate to clean up the cities never quite amassed enough support to overcome the barriers of remote indifference.

Like other chroniclers of urban blight before him, Riis had experimented with illustrations that dramatized the devastating human cost of the tenements. But the line drawings invariably aestheticized the suffering; even the bleakest underground hovel looked almost quaint as an etching. Only photographs seemed to capture the reality with sufficient resolution to change hearts, but whenever Riis experimented with photography, he ran into the same impasse. Almost everything he wanted to photograph involved environments with minimal amounts of light. Indeed, the lack of even indirect sunlight in so many of the tenement flats was part of what made them so objectionable. This was Riis's great stumbling block: as far as photography was concerned, the most important environments in the city – in fact, some of the most important new living quarters in the world – were literally invisible. They couldn't be seen.

All of which should explain Jacob Riis's epiphany at the breakfast table in 1887. Why trifle with line drawings when *Blitzlicht* could shine light in the darkness?

Within two weeks of that breakfast discovery, Riis assembled a team of amateur photographers (and a few curious police officers) to set off into the bowels of the darkened city – literally armed with *Blitzlicht*. (The flash is produced by firing a cartridge of the substance from a revolver.) More than a few denizens of Five Points found the shooting party hard to comprehend. As Riis would later put it: 'The spectacle of half a dozen strange men invading a house in the midnight hour armed with big pistols which they shot off recklessly was hardly reassuring, however sugary our speech, and it was not to be wondered at if the tenants bolted through windows and down fire-escapes wherever we went.'

Before long, Riis replaced the revolver with a frying pan. The apparatus seemed more 'home-like,' he claimed, and made his subjects feel more comfortable encountering the baffling new technology. (The simple act of being photographed was novelty enough for most of them.) It was still dangerous work; one small explosion in the frying pan nearly blinded Riis, and twice he set fire to his house while experimenting with the flash. But the images that emerged from those urban expeditions would ultimately change history. Using new halftone printing techniques, Riis published the photographs in his runaway bestseller, *How the Other Half Lives*, and traveled across the country giving lectures that were accompanied by magic-lantern images of Five Points and its previously invisible poverty. The convention of gathering together in a darkened room and watching illuminated images on a screen would become a ritual of fantasy and wish fulfillment in the twentieth century. But for many Americans, the first images they saw in those environments were ones of squalor and human suffering.

Riis's books and lectures – and the riveting images they contained – helped create a massive shift in public opinion, and set the stage for one of the great periods of social reform in American history. Within a decade of their publication, Riis's images built support for the New York State Tenement House Act of 1901, one of the first great reforms of the Progressive Era, which eliminated much of the appalling living conditions that Riis had documented. His work ignited a new tradition of muckraking that would ultimately improve the working conditions of factory floors as well. In a literal sense, illuminating the dark squalor of the tenements changed the map of urban centers around the world.

Here again we see the strange leaps of the hummingbird's wing at play in social history, new inventions leading to consequences their creators never dreamed of. The utility of mixing magnesium and potassium chlorate seems straightforward enough: *Blitzlicht* meant that human beings could record images in dark environments more accurately than ever before. But that new capability also expanded the space of possibility for other ways of seeing. This is what Riis understood almost immediately. If you could see in the dark, if you could share that vision with strangers around the world thanks to the magic of photography, then the underworld of Five Points could, at long last, be seen in all its tragic reality. The dry, statistical accounts of 'The Report of the Council of Hygiene and Public Health' would be replaced with actual human beings sharing physical space of devastating squalor.

The network of minds that invented flash photography – from the first tinkerers with limelight to Smyth to Miethe and Gaedicke – had deliberately set out with a clearly defined goal: to build a tool that would allow photographs to be taken in darkness. But like almost every important innovation in human history, that breakthrough created a platform that allowed other innovations in radically different fields. We like to organize the world into neat categories: photography goes here, politics there. But the history of *Blitzlicht* reminds us that ideas always travel in networks. They come into being through networks of collaboration, and once unleashed on the world, they set into motion changes that are rarely confined to single disciplines. One century's attempt to invent flash photography transformed the lives of millions of city dwellers in the next century.

Riis's vision should also serve as a corrective to the excesses of crude techno-determinism. It was virtually inevitable that someone would invent flash photography in the nineteenth century. (The simple fact that it was invented multiple times shows us that the time was ripe for the idea.) But there was nothing intrinsic to the technology that suggested it be used to illuminate the lives of the very people who could least afford to enjoy it. You could have reasonably predicted that the problem of photographing in low light would be 'solved' by 1900. But no one would have predicted that its very first mainstream use would come in the form of a crusade against urban poverty. That twist belongs to Riis alone. The march of technology expands the space of possibility around us, but how we explore that space is up to us.

In the fall of 1968, the sixteen members of a graduate studio at the Yale School of Art and Architecture – three faculty and thirteen students – set off on a ten-day expedition to study urban design in the streets of an actual city. This in itself was nothing new: architecture students had been touring the ruins and the monuments of Rome or Paris or Brasília for as long as there have been architecture students. What made this group unusual is that they were leaving behind the Gothic charm of New Haven for a very different kind of city, one that happened to be growing faster than any of the old relics: Las Vegas. It was a city that looked nothing like the dense, concentrated tenements of Riis's Manhattan. But like Riis, the Yale studio sensed that something new and significant was happening on the Vegas strip. Led by Robert Venturi and Denise Scott Brown, the husband-and-wife team who would become founders of postmodern architecture, the Yale studio had been drawn to the desert frontier by the novelty of Vegas, by the shock value they could elicit by taking it seriously, and by the sense that they were watching the future being born. But as much as anything, they had come to Vegas to see a new kind of light. They were drawn, postmodern moths to the flame, to neon.

While neon is technically considered one of the 'rare gases,' it is actually ubiquitous in the earth's atmosphere, just in very small quantities. Each time you take a breath, you are inhaling a tiny amount of neon, mixed in with all the nitrogen and oxygen that saturate breathable air. In the first years of the twentieth century, a French scientist named Georges Claude created a system for liquefying air, which enabled the production of large quantities of liquid nitrogen and oxygen. Processing these elements at industrial scale created an intriguing waste product: neon. Even though neon appears as only one part per 66,000 in ordinary air, Claude's apparatus could produce one hundred liters of neon in a day's work.

With so much neon lying around, Claude decided to see if it was good for anything, and so in proper mad-scientist fashion, he isolated the gas and passed an electrical current through it. Exposed to an electric charge, the gas glowed a vivid shade of red. (The technical term for this process is ionization.) Further experiments revealed that other rare gases such as argon and mercury vapor would produce different colors when electrified, and they were more than five times brighter than conventional incandescent light. Claude quickly patented his neon lights, and set up a display showcasing the invention in front of the Grand Palais in Paris. When demand surged for his product, he established a franchise business for his innovation, not unlike the model employed by McDonald's and Kentucky Fried Chicken years later, and neon lights began to spread across the urban landscapes of Europe and the United States.

In the early 1920s, the electric glow of neon found its way to Tom Young, a British immigrant living in Utah who had started a small business hand-lettering signs. Young recognized that neon could be used for more than just colored light; with the gas enclosed in glass tubes, neon signs could spell out words much more easily than collections of lightbulbs. Licensing Claude's invention, he set up a new business covering the American Southwest. Young realized that the soon-to-be-completed Hoover Dam would bring a vast new source of electricity to the desert, providing a current that could ionize an entire city of neon lights. He formed a new venture, the Young Electric Sign Company, or YESCO. Before long, he found himself building a sign for a new casino and hotel, The Boulders, that was opening in an obscure Nevada town named Las Vegas.

It was a chance collision – a new technology from France finding its way to a sign letterer in Utah – that would create one of the most iconic of twentieth-century urban experiences. Neon advertisements would become a defining feature of big-city centers around the world – think Times Square or Tokyo's Shibuya Crossing. But no city embraced neon with the same unchecked enthusiasm that Las Vegas did, and most of those neon extravaganzas were designed, erected, and maintained by YESCO. 'Las Vegas is the only city in the world whose skyline is made not of buildings . . . but signs,' Tom Wolfe wrote in the middle of the 1960s. 'One can look at Las Vegas from a mile away on route 91 and see no buildings, no trees, only signs. But such signs! They tower. They revolve, they oscillate, they soar in shapes before which the existing vocabulary of art history is helpless.'

It was precisely that helplessness that brought Venturi and Brown to Vegas with their retinue of architecture students in the fall of 1968. Brown and Venturi had sensed that there was a new visual language emerging in that glittering desert oasis, one that didn't fit well with the existing languages of modernist design. To begin with, Vegas had oriented itself around the vantage point of the automobile driver, cruising down Fremont Street or the strip: shop windows and sidewalk displays had given way to sixty-foot neon cowboys. The geometric seriousness of the Seagram Building or Brasília had given way to a playful anarchy: the Wild West of the gold rush thrust up against Olde English feudal designs, sitting next to cartoon arabesques, fronted by an endless stream of wedding chapels. 'Allusion and comment, on the past or present or on our great commonplaces or old clichés, and inclusion of the everyday in the environment, sacred and profane – these are what are lacking in present-day Modern architecture,' Brown and Venturi wrote. 'We can learn about them from Las Vegas as have other artists from their own profane and stylistic sources.'

That language of allusion and comment and cliché was written in neon. Brown and Venturi went so far as to map every single illuminated word visible on Fremont Street. 'In the seventeenth century,' they wrote, 'Rubens created a painting "factory" wherein different workers specialized in drapery, foliage, or nudes. In Las Vegas, there is just such a sign "factory," the Young Electric Sign Company.' Until then, the symbolic frenzy of Vegas had belonged purely to the world of lowbrow commerce: garish signs pointing the way to gambling dens, or worse. But Brown and Venturi had seen something more interesting in all that detritus. As Georges Claude had experienced more than sixty years before, one person's waste is another one's treasure.

Think about these different strands: the atoms of a rare gas, unnoticed until 1898; a scientist and engineer tinkering with the waste product from his 'liquid air'; an enterprising sign designer; and a city blooming implausibly in the desert. All these strands somehow converged to make *Learning from Las Vegas* – a book that architects and urban planners would study and debate for decades – even imaginable as an argument. No other book had as much influence on the postmodern style that would dominate art and architecture over the next two decades.

Learning from Las Vegas gives us a clear case study in how the long-zoom approach reveals elements that are ignored by history's traditional explanatory frameworks: economic or art history, or the 'lone genius' model of innovation. When you ask the question of *why* postmodernism came about as a movement, on some fundamental level the answer has to include Georges Claude and his hundred liters of neon. Claude's innovation wasn't the only cause, by any means, but, in an alternate universe somehow stripped of neon lights, the emergence of postmodern architecture would have in all likelihood followed a different path. The strange interaction between neon gas and electricity, the franchise model of licensing new technology – each served as part of the support structure that made it even possible to conceive of *Learning from Las Vegas*.

This might seem like yet another game of Six Degrees of Kevin Bacon: follow enough chains of causality and you can link postmodernism back to the building of the Great Wall of China or the extinction of the dinosaurs. But the neon-to-postmodernism connections are direct links: Claude creates neon light; Young brings it to Vegas, where Venturi and Brown decide to take its 'revolving and oscillating' glow seriously for the first time. Yes, Venturi and Brown needed electricity, too, but just about everything needed electricity in the 1960s: the moon landing, the Velvet Underground, the 'I Have a Dream' speech. By the same token, Venturi and Brown required the *noble* gases, too; the odds are pretty good that they needed oxygen to write *Learning from Las Vegas*. But it was the rare gas of neon that made their story unique.

Ideas trickle out of science, into the flow of commerce, where they drift into the less predictable eddies of art and philosophy. But sometimes they venture upstream: from aesthetic speculation into hard science. When H. G. Wells published his groundbreaking novel *The War of the Worlds* in 1898, he helped invent the genre of science fiction that would play such a prominent role in the popular imagination during the century that followed. But that book introduced a more specific item to the fledging sci-fi canon: the 'heat ray,' used by the invading Martians to destroy entire towns. 'In some way,' Wells wrote of his technologically savvy aliens, 'they are able to generate an intense heat in a chamber of practically absolute non-conductivity. This intense heat they project in a parallel beam against any object they choose, by means of a polished parabolic mirror of unknown composition, much as the parabolic mirror of a lighthouse projects a beam of light.'

The heat ray was one of those imagined concoctions that somehow get locked into the popular psyche. From *Flash Gordon* to *Star Trek* to *Star Wars*, weapons using concentrated beams of light became almost de rigueur in any sufficiently advanced future civilization. And yet, actual laser beams did not exist until the late 1950s, and didn't become part of everyday life for another two decades after that. Not for the last time, the science-fiction authors were a step or two ahead of the scientists.

But the sci-fi crowd got one thing wrong, at least in the short term. There are no death rays, and the closest thing we have to Flash Gordon's arsenal is laser tag. When lasers did finally enter our lives, they turned out to be lousy for weapons, but brilliant for something the sci-fi authors never imagined: figuring out the cost of a stick of chewing gum.

Like the lightbulb, the laser was not a single invention; instead, as the technology historian Jon Gertner puts it, 'it was the result of a storm of inventions during the 1960s.' Its roots lie in research at Bell Labs and Hughes Aircraft and, most entertainingly, in the independent tinkering of physicist Gordon Gould, who memorably notarized his original design for the laser in a Manhattan candy store, and who went on to have a thirty-year legal battle over the laser patent (a battle he eventually won). A laser is a prodigiously concentrated beam, light's normal chaos reduced down to a single, ordered frequency. 'The laser is to ordinary light,' Bell Lab's John Pierce once remarked, 'as a broadcast signal is to static.'

Unlike the lightbulb, however, the early interest in the laser was not motivated by a clear vision of a consumer product. Researchers knew that the concentrated signal of the laser could be used to embed information more efficiently than could existing electrical wiring, but exactly how that bandwidth would be put to use was less evident. 'When something as closely related to signaling and communication as this comes along,' Pierce explained at the time, 'and something is new and little understood, and you have the people who can do something about it, you'd just better do it, and worry later just about the details of why you went into it.' Eventually, as we have already seen, laser technology would prove crucial to digital communications, thanks to its role in fiber optics. But the laser's first critical application would appear at the checkout counter, with the emergence of bar-code scanners in the mid-1970s.

The idea of creating some kind of machine-readable code to identify products and prices had been floating around for nearly half a century. Inspired by the dashes and dots of Morse code, an inventor named Norman Joseph Woodland designed a visual code that resembled a bull's-eye in the 1950s, but it required a five-*hundred*-watt bulb – almost ten times brighter than your average lightbulb – to read the code, and even then it wasn't very accurate. Scanning a series of black-and-white symbols turned out to be the kind of job that lasers immediately excelled at, even in their infancy. By the early 1970s, just a few years after the first working lasers debuted, the modern system of bar codes – known as the Universal Product Code – emerged as the dominant standard. On June 26, 1974, a stick of chewing gum in a supermarket in Ohio became the first product in history to have its bar code scanned by a laser. The technology spread slowly: only one percent of stores had bar-code scanners as late as 1978. But today, almost everything you can buy has a bar code on it.

In 2012, an economics professor named Emek Basker published a paper that assessed the impact of bar-code scanning on the economy, documenting the spread of the technology through both mom-and-pop stores and big chains. Basker's data confirmed the classic trade-offs of early adoption: most stores that integrated bar-code scanners in the early years didn't see much benefit from them, since employees had to be trained to use the new technology, and many products didn't have bar codes yet. Over time, however, the productivity gains became substantial, as bar codes became ubiquitous. But the most striking discovery in Basker's research was this: The productivity gains from bar-code scanners were not evenly distributed. Big stores did much better than small stores.

There have always been inherent advantages to maintaining a large inventory of items in a store: the customer has more options to choose from, and items can be purchased in bulk from wholesalers for less money. But in the days before bar codes and other forms of computerized inventory-management tools, the benefits of housing a vast inventory were largely offset by the costs of keeping track of everything. If you kept a thousand items in stock instead of a hundred, you needed more people and time to figure out which sought-after items needed restocking and which were just sitting on the shelves taking up space. But bar codes and scanners greatly reduced the costs of maintaining a large inventory. The decades after the introduction of the bar-code scanner in the United States witnessed an explosion in the size of retail stores; with automated inventory management, chains were free to balloon into the epic big-box stores that now dominate retail shopping. Without bar-code scanning, the modern shopping landscape of Target and Best Buy and supermarkets the size of airport terminals would have had a much harder time coming into being. If there was a death ray in the history of the laser, it was the metaphoric one directed at the mom-and-pop, indie stores demolished by the big-box revolution.

While the early sci-fi fans of *War of the Worlds* and *Flash Gordon* would be disappointed to see the mighty laser scanning packets of chewing gum – its brilliantly concentrated light harnessed for inventory management – their spirits would likely improve contemplating the National Ignition Facility, at the Lawrence Livermore Labs in Northern California, where scientists have built the world's largest and highest-energy laser system. Artificial light began as simple illumination, helping us read and entertain ourselves after dark; before long it had been transformed into advertising and art and information. But at NIF, they are taking light full circle, using lasers to create a new source of energy based on nuclear fusion, re-creating the process that occurs naturally in the dense core of the sun, our original source of natural light.

Deep inside the NIF, near the 'target chamber,' where the fusion takes place, a long hallway is decorated with what appears, at first glance, to be a series of identical Rothko paintings, each displaying eight large red squares the size of a dinner plate. There are 192 of them in total, each representing one of the lasers that simultaneously fire on a tiny bead of hydrogen in the ignition chamber. We are used to seeing lasers as a pinpoint of concentrated light, but at NIF, the lasers are more like cannonballs, almost two hundred of them summed together to create a beam of energy that would have made H. G. Wells proud.

The multibillion-dollar complex has all been engineered to execute discrete, microsecond-long events: firing the lasers at the hydrogen fuel while hundreds of sensors and high-speed cameras observe the activity. Inside the NIF, they refer to these events as 'shots.' Each shot requires the meticulous orchestration of more than six hundred thousand controls. Each laser beam travels 1.5 kilometers guided by a series of lenses and mirrors, and combined they build in power until they reach 1.8 million joules of energy and five-hundred-trillion watts, all converging on a fuel source the size of a peppercorn. The lasers have to be positioned with a breathtaking accuracy, the equivalent of standing on the pitcher's mound at AT&T Park in San Francisco and throwing a strike at Dodger Stadium in Los Angeles, some 350 miles away. Each microsecond pulse of light has, for its brief existence, a thousand times the amount of energy in America's entire national grid.

When all of NIF's energy slams into its millimeter-sized targets, unprecedented conditions are generated in the target materials – temperatures of more than a hundred million degrees, densities up to a hundred times the density of lead, and pressures more than a hundred billion times Earth's atmospheric pressure. These conditions are similar to those inside stars, the cores of giant planets, and nuclear weapons – allowing NIF to create, in essence, a miniature star on Earth, fusing hydrogen atoms together and releasing a staggering amount of energy. For that fleeting moment, as the lasers compress the hydrogen, that fuel pellet is the hottest place in the solar system – hotter, even, than the core of the sun.

The goal of the NIF is not to create a death ray – or the ultimate bar-code scanner. The goal is to create a sustainable source of clean energy. In 2013, NIF announced that the device had for the first time generated net positive energy during several of its shots; by a slender margin, the fusion process required less energy than it created. It is still not enough to reproduce efficiently on a mass scale, but NIF scientists believe that with enough experimentation, they will eventually be able to use their lasers to compress the fuel pellet with almost perfect symmetry. At that point, we would have a potentially limitless source of energy to power all the lightbulbs and neon signs and bar-code scanners – not to mention computers and air-conditioners and electric cars – that modern life depends on.

Those 192 lasers converging on the hydrogen pellet are a telling reminder of how far we have come in a remarkably short amount of time. Just two hundred years ago, the most advanced form of artificial light involved cutting up a whale on the deck of a boat in the middle of the ocean. Today we can use light to create an artificial sun on Earth, if only for a split-second. No one knows if the NIF scientists will reach their goal of a clean, sustainable fusion-based energy source. Some might even see it as a fool's errand, a glorified laser show that will never return more energy than it takes in. But setting off for a three-year voyage into the middle of the Pacific Ocean in search of eighty-foot sea mammals was every bit as crazy, and somehow that quest fueled our appetite for light for a century. Perhaps the visionaries at NIF – or another team of muckers somewhere in the world – will eventually do the same. One way or another, we are still chasing new light.

CONCLUSION

The Time Travelers

On July 8, 1835, an English baron by the name of William King was married in a small ceremony in the western suburbs of London, at an estate called Fordhook that had once belonged to the novelist Henry Fielding. By all accounts it was a pleasant wedding, though it was a much smaller affair than one might have expected given King's title and family wealth. The intimacy of the wedding was due to the general public's fascination with his nineteen-year-old bride, the beautiful and brilliant Augusta Byron, now commonly known by her middle name of Ada, daughter of the notorious Romantic poet Lord Byron. Byron had been dead for a decade, and had not seen his daughter since she was an infant, but his reputation for creative brilliance and moral dissolution continued to reverberate through European culture. There were no paparazzi to hound Baron King and his bride in 1835, but Ada's fame meant that a certain measure of discretion was required at her wedding.

After a short honeymoon, Ada and her new husband began dividing their time between his family estate in Ockham, another estate in Somerset, and a London home, beginning what promised to be a life of domestic leisure, albeit challenged by the enviable difficulties of maintaining three residences. By 1840, the couple had produced three children, and King had been elevated to earldom with Queen Victoria's coronation list.

By the conventional standards of Victorian society, Ada's life would have seemed any woman's dream: nobility, a loving husband, and three children, including the all-important male heir. But as she settled into the duties of motherhood and of running a landed estate, she found herself fraying at the edges, drawn to paths that were effectively unheard-of for Victorian women. In the 1840s, it was not outside the bounds of possibility for a woman to be engaged in the creative arts in some fashion, and even to dabble in writing her own fiction or essays. But Ada's mind was drawn in another direction. She had a passion for numbers.

When Ada was a teenager, her mother, Annabella Byron, had encouraged her study of mathematics, hiring a series of tutors to instruct her in algebra and trigonometry, a radical course of study in an age when women were excluded from important scientific institutions such as the Royal Society, and were assumed to be incapable of rigorous scientific thinking. But Annabella had an ulterior motive in encouraging her daughter's math skills, hoping that the methodical and practical nature of her studies would override the dangerous influence of her dead father. A world of numbers, Annabella hoped, would save her daughter from the debauchery of art.

For a time, it appeared that Annabella's plan had worked. Ada's husband had been made Earl of Lovelace, and as a family they seemed on a path to avoid the chaos and unconventionality that had destroyed Lord Byron fifteen years before. But as her third child grew out of infancy, Ada found herself drawn back to the world of math, feeling unfulfilled by the domestic responsibilities of Victorian motherhood. Her letters from the period display a strange mix of Romantic ambition – the sense of a soul larger than the ordinary reality it has found itself trapped in – combined with an intense belief in the power of mathematical reason. Ada wrote about differential calculus with the same passion and exuberance (and self-confidence) that her father wrote about forbidden love:

> Owing to some peculiarity in my nervous system, I have perceptions of some things, which no one else has . . . an intuitive perception of hidden things; – that is of things hidden away from eyes, ears, and the ordinary senses. This alone would advantage me little, in the discovery line, but there is, secondly, my immense reasoning faculties, and my concentrative faculty.

In the late months of 1841, Ada's conflicted feelings about her domestic life and her mathematical ambitions came to a crisis point, when she learned from Annabella that, in the years before his death, Lord Byron had conceived a daughter with his half sister. Ada's father was not only the most notorious author of his time; he was also guilty of incest, and the offspring of this scandalous union was a girl Ada had known for many years. Annabella had volunteered the news to her daughter as definitive proof that Byron was a wretch, and that such a rebellious, unconventional lifestyle could only end in ruin.

And so, at the still young age of twenty-five, Ada Lovelace found herself at a crossroads, confronting two very different ways of being an adult in the world. She could resign herself to the settled path of a baroness and live within the boundaries of conventional decorum. Or she could embrace those 'peculiarities of [her] nervous system' and seek out some original path for herself and her distinctive gifts.

It was a choice that was deeply situated in the culture of Ada's time: the assumptions that framed and delimited the roles women could adopt, the inherited wealth that gave her the option of choosing in the first place, and the leisure time to mull over the decision. But the paths in front of her were also carved out by her genes, by the talents and dispositions – even the mania – Ada had inherited from her parents. In choosing between domestic stability and some unknown break from convention, she was, in a sense, choosing between her mother and her father. To stay settled at Ockham Park was the easier path; all the forces of society propelled her toward it. And yet, like it or not, she was still Byron's daughter. A conventional life seemed increasingly unthinkable.

But Ada Lovelace found a way around the impasse she had confronted in her mid-twenties. In collaboration with another brilliant Victorian who was equally ahead of his time, Ada charted a path that allowed her to push the barriers of Victorian society without succumbing to the creative chaos that had enveloped her father. She became a software programmer.

Writing code in the middle of the nineteenth century may seem like a vocation that would be possible only with time travel, but as chance would have it, Ada had met the one Victorian who was capable of giving her such a project: Charles Babbage, the brilliant and eclectic inventor who was in the middle of drafting plans for his visionary Analytical Engine. Babbage had spent the previous two decades concocting state-of-the-art calculators, but since the mid-1830s, he had commenced work on a project that would last the rest of his life: designing a truly programmable computer, capable of executing complex sequences of calculations that went far beyond the capabilities of any contemporary machines. Babbage's Analytical Engine was doomed to a certain practical failure – he was trying to build a digital-age computer with industrial-age mechanical parts – but conceptually it was a brilliant leap forward. Babbage's design anticipated all the major components of modern computers: the notion of a central processing unit (which Babbage dubbed 'the mill'), of random-access memory, and of software that would control the machine, etched on the very same punch cards that would be used to program computers more than a century later.

Ada had met Babbage at the age of seventeen, in one of his celebrated London salons, and the two had kept up a friendly and intellectually lively correspondence over the years. And so when she hit her crossroads in the early 1840s, she wrote a letter to Babbage that suggested he might prove to be a potential escape route from the limitations of life at Ockham Park:

> I am very anxious to talk to you. I will give you a hint on what. It strikes me that at some future time, my head may be

> made by you subservient to some of your purposes & plans. If so, if ever I could be worth or capable of being used by you, my head will be yours.

Babbage, as it turned out, did have a use for Ada's remarkable head, and their collaboration would lead to one of the founding documents in the history of computing. An Italian engineer had written an essay on Babbage's machine, and at the recommendation of a friend, Ada translated the text into English. When she told Babbage of her work, he asked why she hadn't written her own essay on the subject. Despite all her ambition, the thought of composing her own analysis had apparently never occurred to Ada, and so at Babbage's encouragement, she concocted her own aphoristic commentary, stitched together out of a series of extended footnotes attached to the Italian paper.

Those footnotes would ultimately prove to be far more valuable and influential than the original text they annotated. They contained a series of elemental instruction sets that could be used to direct the calculations of the Analytical Engine. These are now considered to be the first examples of working software ever published, though the machinery that could actually run the code wouldn't be built for another century.

There is some dispute over whether Ada was the sole author of these programs, or whether she was refining routines that Babbage himself had worked out previously. But Ada's greatest contribution lay not in writing out instruction sets, but rather in envisioning a range of utility for the machine that Babbage himself had not considered. 'Many persons,' she wrote, 'imagine that because the business of the engine is to give its results in numerical notation the nature of its processes must consequently be arithmetical and numerical, rather than algebraical and analytical. This is an error. The engine can arrange and combine its numerical quantities exactly as if they were letters or any other general symbols.' Ada recognized that Babbage's machine was not a mere number cruncher. Its potential uses went far beyond rote calculation. It might even someday be capable of the higher arts:

> Supposing, for instance, that the fundamental relations of pitched sounds in the science of harmony and musical composition were susceptible of such expressions and adaptations, the Engine might compose elaborate and scientific pieces of music of any degree of complexity or extent.

To have this imaginative leap in the middle of the nineteenth century is almost beyond comprehension. It was hard enough to wrap one's mind around the idea of programmable computers – almost all of Babbage's contemporaries failed to grasp what he had invented – but somehow, Ada was able to take the concept one step further, to the idea that this machine might also conjure up language and art. That one footnote opened up a conceptual space that would eventually be occupied by so much of early twenty-first-century culture: Google queries, electronic music, iTunes, hypertext. The computer would not just be an unusually flexible calculator; it would be an expressive, representational, even aesthetic machine.

Of course, Babbage's idea and Lovelace's footnote proved to be so far ahead of their time that, for a long while, they were lost to history. Most of Babbage's core insights had to be independently rediscovered a hundred years later, when the first working computers were built in the 1940s, running on electricity and vacuum tubes instead of steam power. The notion of computers as aesthetic tools, capable of producing culture as well as calculation, didn't become widespread – even in high-tech hubs such as Boston or Silicon Valley – until the 1970s.

Most important innovations – in modern times at least – arrive in clusters of simultaneous discovery. The conceptual and technological pieces come together to make a certain idea imaginable – artificial refrigeration, say, or the lightbulb – and all around the world, you suddenly see people working on the problem, and usually approaching it with the same fundamental assumptions about how that problem is ultimately going to be solved. Edison and his peers may have disagreed about the importance of the vacuum or the carbon filament in inventing the electric lightbulb, but none of them were working on an LED. The predominance of simultaneous, multiple invention in the historical record has interesting implications for the philosophy of history and science: To what extent is the *sequence* of invention set in stone by the basic laws of physics or information or the biological and chemical constraints of the earth's environment? We take it for granted that microwaves have to be invented after the mastery of fire, but how inevitable is it that, say, telescopes and microscopes quickly followed the invention of spectacles? (Could one imagine, for instance, spectacles being widely adopted, but then a pause of five hundred years before someone thinks of rejiggering them into a telescope? It seems unlikely, but I suppose it's not impossible.) The fact that these simultaneous-invention clusters are so pronounced in the fossil record of technology tells us, at the very least, that some confluence of historical events has made a new technology imaginable in a way that it wasn't before.

What those events happen to be is a murkier but fascinating question. I have tried to sketch a few answers here. Lenses, for instance, emerged out of several distinct developments: glassmaking expertise, particularly as cultivated on Murano; the adoption of glass 'orbs' that helped monks read their scrolls later in life; the invention of the printing press, which created a surge in demand for spectacles. (And, of course, the basic physical properties of silicon dioxide itself.) We can't know for certain the full extent of these influences, and no doubt some influences are too subtle for us to detect after so many years, like starlight from remote suns. But the question is nonetheless worth exploring, even if we are resigned to somewhat speculative answers, the same way we are when we try to wrestle with the causes behind the American Civil War or the droughts of the Dust Bowl era. They're worth exploring because we are living through comparable revolutions today, set by the boundaries and opportunities of our own adjacent possible. Learning from the patterns of innovation that shaped society in the past can only help us navigate the future more successfully, even if our explanations of that past are not falsifiable in quite the same way that a scientific theory is.

*

But if simultaneous invention is the rule, what about the exceptions? What about Babbage and Lovelace, who were effectively a century ahead of just about every other human being on the planet? Most innovation happens in the present tense of the adjacent possible, working with the tools and concepts that are available in that time. But every now and then, some individual or group makes a leap that seems almost like time traveling. How do they do it? What allows them to see past the boundaries of the adjacent possible when their contemporaries fail to do so? That may be the greatest mystery of all.

The conventional explanation is the all-purpose but somewhat circular category of 'genius.' Da Vinci could imagine (and draw) helicopters in the fifteenth century because he was a genius; Babbage and Lovelace could imagine programmable computers in the nineteenth century because they were geniuses. No doubt all three were blessed with great intellectual gifts, but history is replete with high-IQ individuals who don't manage to come up with inventions that are decades or centuries ahead of their time. Some of that time-traveling genius no doubt came from their raw intellectual skills, but I suspect just as much came out of the environment their ideas evolved in, the network of interests and influence that shaped their thinking.

If there is a common thread to the time travelers, beyond the nonexplanation of genius, it is this: they worked at the margins of their official fields, or at the intersection point between very different disciplines. Think of Édouard-Léon Scott de Martinville inventing his sound-recording device a generation before Edison began working on the phonograph. Scott was able to imagine the idea of 'writing' sound waves because he had borrowed metaphors from stenography and printing and anatomical studies of the human ear. Ada Lovelace could see the aesthetic possibilities of Babbage's Analytical Engine because her life had been lived at a unique collision point between advanced math and Romantic poetry. The 'peculiarities' of her 'nervous system' – that Romantic instinct to see beyond the surface appearances of things – allowed her to imagine a machine capable of manipulating symbols or composing music, in a way that even Babbage himself had failed to do.

To a certain extent, the time travelers remind us that working within an established field is both empowering and restricting at the same time. Stay within the boundaries of your discipline, and you will have an easier time making incremental improvements, opening the doors of the adjacent possible that are directly available to you given the specifics of the historical moment. (There's nothing wrong with that, of course. Progress depends on incremental improvements.) But those disciplinary boundaries can also serve as blinders, keeping you from the bigger idea that becomes visible only when you cross those borders. Sometimes those borders are literal, geographic ones: Frederic Tudor traveling to the Caribbean and dreaming of ice in the tropics; Clarence Birdseye ice fishing with the Inuits in Labrador. Sometimes the boundaries are conceptual: Scott borrowing the metaphors of stenography to invent the phonautograph. The time travelers tend, as a group, to have a lot of hobbies: think of Darwin and his orchids. When Darwin published his book on pollination four years after *Origin of*

Species, he gave it the wonderfully Victorian title, *On the Various Contrivances by Which British and Foreign Orchids are Fertilised by Insects, and on the Good Effects of Intercrossing*. We now understand the 'good effects of intercrossing' thanks to the modern science of genetics, but the principle applies to intellectual history as well. The time travelers are unusually adept at 'intercrossing' different fields of expertise. That's the beauty of the hobbyist: it's generally easier to mix different intellectual fields when you have a whole array of them littering your study or your garage.

One of the reasons garages have become such an emblem of the innovator's workspace is precisely because they exist outside the traditional spaces of work or research. They are not office cubicles or university labs; they're places away from work and school, places where our peripheral interests have the room to grow and evolve. Experts head off to their corner offices and lecture halls. The garage is the space for the hacker, the tinkerer, the maker. The garage is not defined by a single field or industry; instead, it is defined by the eclectic interests of its inhabitants. It is a space where intellectual networks converge.

In his famous Stanford commencement speech, Steve Jobs – the great garage innovator of our time – told several stories about the creative power of stumbling into new experiences: dropping out of college and sitting in on a calligraphy class that would ultimately shape the graphic interface of the Macintosh; being forced out of Apple at the age of thirty, which enabled him to launch Pixar into animated movies and create the NeXT computer. 'The heaviness of being successful,' Jobs explained, 'was replaced by the lightness of being a beginner again, less sure about everything. It freed me to enter one of the most creative periods of my life.'

Yet there is a strange irony at the end of Jobs's speech. After documenting the ways that unlikely collisions and explorations can liberate the mind, he ended with a more sentimental appeal to be 'true to yourself':

> Don't be trapped by dogma – which is living with the results of other people's thinking. Don't let the noise of others' opinions drown out your own inner voice. And most important, have the courage to follow your heart and intuition.

If there's anything we know from the history of innovation – and particularly from the history of the time travelers – it is that being true to yourself is not enough. Certainly, you don't want to be trapped by orthodoxy and conventional wisdom. Certainly, the innovators profiled in this book had the tenacity to stick with their hunches for long periods of time. But there is comparable risk in being true to your own sense of identity, your own roots. Better to challenge those intuitions, explore uncharted terrain, both literal and figurative. Better to make new connections than remain comfortably situated in the same routine. If you want to improve the world slightly, you need focus and determination; you need to stay within the confines of a field and open the new doors in the adjacent possible one at a time. But if you want to be like Ada, if you want to have an 'intuitive perception of hidden things' – well, in that case, you need to get a little lost.

Acknowledgments

There is a predictable social rhythm to writing books, in my experience at least: they begin very close to solitude, the writer alone with his or her ideas, and they stay in that intimate space for months, sometimes years, interrupted only by the occasional interview or conversation with an editor. And then, as publication nears, the circle widens: suddenly a dozen people are reading and helping usher a rough, unformed manuscript into life as a polished final product. And then the book hits the shelves, and all that work becomes almost terrifyingly public, with thousands of bookstore employees, reviewers, radio interviewers, and readers interacting with words that began their life in such a private embrace. And then the whole cycle starts all over again.

But this book followed a completely different pattern. It was a social, collaborative process from the very beginning, thanks to the simultaneous development of our PBS/BBC television series. The stories and observations – not to mention the overarching structure of the book – evolved out of hundreds of conversations: in California and London and New York and Washington, via e-mail and Skype, with dozens of people. Making the series and book was the hardest work I have ever done in my life – and not just when they forced me to descend into the sewers of San Francisco. But it was also the most rewarding work I've ever done, in large part because my collaborators were such inventive and entertaining people. This book has benefited from their intelligence and support in a thousand different ways.

My gratitude begins with the irrepressible Jane Root, who persuaded me to try my hand at television, and remained a tireless champion of this project throughout its life. (Thanks to Michael Jackson for introducing us so many years ago.) As producers, Peter Lovering, Phil Craig, and Diene Petterle shaped the ideas and narratives in this book with great skill and creativity, as did the directors Julian Jones, Paul Olding, and Nic Stacey. A project this complex, with so many potential narrative threads, would have been almost impossible to complete without the help of our researchers and story producers, Jemila Twinch, Simon Willgoss, Rowan Greenaway, Robert MacAndrew, Gemma Hagen, Jack Chapman, Jez Bradshaw, and Miriam Reeves. I'd also like to thank Helena Tait, Kirsty Urquhart-Davies, Jenny Wolf, and the rest of the team at Nutopia. (Not to mention the brilliant illustrators at Peepshow Collective.) At PBS/OPB, I'm indebted to the extraordinary vote of confidence from Beth Hoppe, Bill Gardner, Dave Davis, and Jennifer Lawson from CPB, along with Martin Davidson at the BBC.

A book that covers so many different fields can only succeed by drawing on the expertise of others. I'm grateful to the many talented people we interviewed for this project, some of whom were kind enough to read portions of the manuscript in draft: Terri Adams, Katherine Ashenburg, Rosa Barovier, Stewart Brand, Jason Brown, Dr Ray Briggs, Stan Bunger, Kevin Connor, Gene Chruszcs, John DeGenova, Jason Deichler, Jacques Desbois, Dr Mike Dunne, Caterina Fake, Kevin Fitzpatrick, Gai Gherardi,

David Giovannoni, Peggi Godwin, Thomas Goetz, Alvin Hall, Grant Hill, Sharon Hudgens, Kevin Kelly, Craig Koslofsky, Alan MacFarlane, David Marshall, Demetrios Matsakis, Alexis McCrossen, Holley Muraco, Lyndon Murray, Bernard Nagengast, Max Nova, Mark Osterman, Blair Perkins, Lawrence Pettinelli, Dr Rachel Rampy, Iegor Reznikoff, Eamon Ryan, Jennifer Ryan, Michael D. Ryan, Steven Ruzin, Davide Salvatore, Tom Scheffer, Eric B. Schultz, Emily Thompson, Jerri Thrasher, Bill Wasik, Jeff Young Ed Yong; and Carl Zimmer.

At Riverhead, my editor and publisher Geoffrey Kloske's usual astute sense of what the book needed editorially was accompanied by an artful vision of the book's design that shaped the project from the very beginning. Thanks also to Casey Blue James, Hal Fessenden, and Kate Stark at Riverhead, and my UK publishers, Stefan McGrath and Josephine Greywoode. As always, thanks to my agent, Lydia Wills, for keeping faith in this project for almost half a decade.

Finally, my love and gratitude to my wife, Alexa, and my sons, Clay, Rowan, and Dean. Writing books for a living has generally meant that I spend more time with them, procrastinating by puttering around the house and chatting with Alexa, picking the kids up from school. But this project took me away from home more than it kept me there. So thanks to all four of you for tolerating my absences. Hopefully they made the heart grow fonder. I know they did mine.

Notes

Introduction

1–2 'We could imagine' . . . 'system of cogs and wheels': De Landa, p. 3.

11 'I have a friend who's an artist': From *The Pleasure of Finding Things Out*, a 1981 documentary.

Chapter 1. Glass

15 A small community of glassmakers from Turkey: Willach, p. 30.

16 In 1291, in an effort: Toso, p. 34.

16 After years of trial and error . . . Angelo Barovier: Verità, p. 63.

18 For several generations, these ingenious new devices: Dreyfus, pp. 93–106.

19 Within a hundred years of Gutenberg's invention: http://faao.org/what/heritage/exhibits/online/spectacles/.

20 Legend has it that one of them: Pendergrast, p. 86.

22 'one of the worst teachers': Quoted in Hecht, p. 30.

23 'If I had been promised': Quoted ibid., p. 31.

26 Some of the most revered works of art: Woods-Marsden, p. 31.

26 Back in Murano, glassmakers had figured out: Pendergrast, pp. 119–120.

28 'When you wish to see': Quoted ibid., p. 138.

28 'It is as if all humans': Macfarlane and Martin, p. 69.

28 'The most powerful prince in the world': Mumford, p. 129.

35 'How from these ashes': Quoted ibid., p. 131.

Chapter 2. Cold

39 'Ice is an interesting subject': Thoreau, p. 192.

41 'Plan etc for transporting Ice to Tropical Climates': Quoted in Weightman, loc. 274–276.

42 'In a country where at some seasons': Quoted ibid., loc. 289–290.

42 'fortunes larger than we shall know what to do with': Quoted ibid., loc. 330.

42–43 'No joke. A vessel': Quoted ibid., loc. 462–463.

45 'On Monday the 9th instant': Quoted ibid., loc. 684–688.

48 'This day I sailed from Boston': Quoted ibid., loc. 1911–1913.

49 'Thus it appears that the sweltering inhabitants': Thoreau, p. 193.

50 'In workshops, composing rooms, counting houses': Quoted in Weightman, loc. 2620–2621.

52 'cooling rooms packed with natural ice': Miller, p. 205.

52 'It was this application of elementary physics': Ibid., p. 208.

53 'a city-country [food] system that was the most powerful': Ibid.

53 'the greatest aggregation of labor': Sinclair.

53 'a direct sloping path': Dreiser, p. 620.

55 A string of shipwrecks delayed ice shipments: Wright, p. 12.

58 'might better serve mankind': Quoted in Gladstone, p. 34.

60 By 1870, the southern states: Shachtman, p. 75.

61 Any meat or produce that had been frozen: Kurlansky, pp. 39–40.

63 'The inefficiency and lack of sanitation': Quoted ibid., p. 129.

66 His first great test came: www.filmjournal.com/filmjournal/content_display/news-and-features/features/technology/e3iad1c03f082a43aa277a9bb65d3d561b5.

67 'It takes time to pull down': Ingels, p. 67.

69 Swelling populations in Florida, Texas: Polsby, pp. 80–88.

71 Millions of human beings around the world: www.theguardian.com/society/2013/jul/12/story-ivf-five-million-babies.

Chapter 3. Sound

74 Reznikoff's theory is that Neanderthal communities: http://www.musicandmeaning.net/issues/showArticle.php?artID=3.2.

77 In the annals of invention . . . Phonautograph: Klooster, p. 263.

79 Just a few years ago, a team of sound historians: www.firstsounds.org.

81 His name was Alexander Graham Bell: Mercer, pp. 31–32.

83 'It may sound ridiculous to say': Quoted in Gleick 2012, loc. 3251–3257.

84 Eventually, the antitrust lawyers: Gertner, pp. 270–271.

86 Effectively, they were taking snapshots: http://www.nsa.gov/about/cryptologic_heritage/center_crypt_history/publications/sigsaly_start_digital.shtml.

87 'We are assembled today': Quoted ibid.

89 Working out of his home lab: Hijiya, p. 58.

89 As a transmission device for the spoken word: Thompson, p. 92.

90 'I look forward to the day': Quoted in Fang, p. 93.

90 'The ether wave passing over the tallest towers': Quoted in Adams, p. 106.

91 But somehow, lurking behind all of De Forest's accumulation: Hijiya, p. 77.

92 Almost overnight, radio made jazz: Carney, pp. 36–37.

94 'It is no wonder that so much of the search for': Quoted in Brown, p. 176.

96 'Sympathetic to the society's mission': Thompson, pp. 148–158.

97 'No one could figure out the sound': Quoted in Diekman, p. 75.

100 Just a few days before the sinking: Frost, p. 466.

102 The German U-boats roaming the North Atlantic: Ibid., p. 476–477.

102 'I pleaded with them': Quoted ibid., p. 478.

104 China was almost 110 boys: Yi, p. 294.

Chapter 4. Clean

107 In December 1856, a middle-aged Chicago engineer: Cain, p. 355.

107 During the Pleistocene era, vast ice fields: Miller, p. 68.

108 'You have been guilty': Quoted ibid., p. 70.

108 'green and black slime': Miller, p. 75.

109 That rate of growth ... a lot of excrement: Chesbrough, 1871.

109 'The gutters are running': Quoted in Miller, p. 123.

109 'The river is positively red': Quoted ibid., p. 123.

109–110 Many of them subscribed . . .'death fogs': Miller, p. 123.

110 'the most competent engineer': Cain, p. 356.

111 Building by building, Chicago was lifted: Ibid., p. 357.

111 'The people were in [the hotel]': Cohn, p. 16.

112 'Never a day passed': Macrae, p. 191.

112 Within three decades, more than twenty cities: Burian, Nix, Pitt, and Durrans.

113 'came out cooked': www.pbs.org/wgbh/amex/chicago/peopleevents/e_canal.html.

113 'The grease and chemicals': Sinclair, p. 110.

115 Working in Vienna's General Hospital: Goetz, loc. 612–615.

116 'Bathing fills the head': Quoted in Ashenburg, p. 100.

116 As a child, Louis XIII: Ashenburg, p. 105.

117 Harriet Beecher Stowe and her sister: Ibid., p. 221.

117 'By the last decades': Ibid., p. 201.

119 'A large part of my success': www.zeiss.com/microscopy/en_us/about-us/nobel-prize-winners.html.

119 Koch established a unit of measure: McGuire, p. 50.

120 It was an interest born: Ibid., pp. 112–113.

122 'Leal did not have time': Ibid., p. 200.

123 'I do there find and report': Quoted in ibid., p. 248.

124 'And if the experiment turned out': Quoted ibid., p. 228.

124 About a decade ago, two Harvard professors: Cutler and Miller, pp. 1–22.

125 'In total, a woman's thighs': Wiltse, p. 112.

127 Annie Murray had created America's first commercial bleach: *The Clorox Company: 100* Years, *1,000* Reasons (The Clorox Company, 2013), pp. 18–22.

131 In 2012, the Bill and Melinda Gates Foundation: http://www.gatesfoundation.org/What-We-Do/Global-Development/Reinvent-the-Toilet-Challenge.

Chapter 5. Time

135 In October 1967, a group of scientists from around the world . . . But the General Conference on Weights and Measures: Blair, p. 246.

137 To confirm his observations: Kreitzman, p. 33.

138 'The marvelous property of the pendulum': Drake, loc. 1639.

139 His astronomical observations had suggested: www://galileo.rice.edu/sci/instruments/pendulum.html.

141 The watchmakers were the advance guard: Mumford, p. 134.

142 'On a rainy day': Thompson, pp. 71–72.

143 'the employer must use the time of his labour': Ibid., p. 61.

143 'deadly statistical clock': Dickens, p. 130.

145 Dennison had a vision of machines: Priestley, p. 5.

145 Dennison's 'Wm. Ellery' watch . . . cost just $3.50: Ibid., p. 21.

148 'It is simply preposterous': www://srnteach.us/HIST1700/assets/projects/unit3/docs/railroads.pdf.

148 The United States remained temporally challenged . . . William F. Allen: McCrossen, p. 92.

148 'the day of two noons': Bartky, pp. 41–42.

149 Instead, pulses of electricity traveling: McCrossen, p. 107.

155 In the 1890s . . . Marie Curie proposed: Senior, pp. 244–245.

157 'a clock that ticks once a year': http://longnow.org/clock/.

158 'If you have a Clock ticking': Ibid.

Chapter 6. Light

163 In a diary entry from 1743: Irwin, p. 47.

163 When darkness fell, they would drift: Ekirch, p. 306.

164 In the deep waters of the North Atlantic: Dolin, loc. 1272.

165 In a 1751 letter, Ben Franklin: Quoted ibid., loc. 1969–1971.

165 The candle business became so lucrative: Dolin, loc. 1992.

166 It's remarkable to think: Irwin, p. 50.

166 Somewhere on the order of three hundred thousand: Ibid., pp. 51–52.

167 'During periods of major technological change': Nordhaus, p. 29.

168 Today, you can buy three hundred days of artificial light: Ibid., p. 37.

169 The problem with this story: Friedel, Israel, and Finn, loc. 1475.

173 'celluloid, wood shavings': Ibid., loc. 1317–1320.

175 'I cannot help laughing': Quoted in Stross, loc. 1614.

176 What Edison and the muckers created: Friedel, Israel, and Finn, loc. 2637.

178 Smyth interpreted this correspondence: Bruck, p. 104.

180 In October 1887, a New York paper . . . Blitzlicht: Riis, loc. 2228.

181 'We used to go in the small hours': Ibid., loc. 2226.

183 'The spectacle of half a dozen': Ibid., loc. 2238.

184 Within a decade of their publication: Yochelson, p. 148.

187 Even though neon appears: Ribbat, pp. 31–33.

188 In the early 1920s, the electric glow: Ibid., pp. 82–83.

188 'Las Vegas is the only city': Wolfe, p. 7.

189 'Allusion and comment, on the past or present': Venturi, Scott Brown, and Izenour, p. 21.

191 'In some way they are able to generate': Wells, p. 28.

192 'it was the result of a storm of inventions': Gertner, p. 256.

192 'The laser is to ordinary light': Ibid., p. 255.

194 Big stores did much better: Basker, pp. 21–23.

Conclusion: The Time Travelers

200 A world of numbers: Toole, p. 20.

201 'Owing to some peculiarity': Quoted in Swade, p. 158.

203 'I am very anxious to talk to you': Quoted ibid., p. 159.

205 'Supposing, for instance, that the fundamental': Quoted ibid., p. 170.

Bibliography

Adams, Mike. *Lee de Forest: King of Radio, Television and Film.* Springer/ Copernicus Books, 2012.

Allen, William F. 'Report on the subject of National Standard Time Made to the General and Southern Railway Time Convention held in St Louis, April 11, 1883, and in New York City, April 18, 1883.' New York Public Library. http://archives.nypl.org/uploads/collection/pdf_finding_aid/allenwf.pdf

Ashenburg, Katherine. *The Dirt on Clean: An Unsanitized History.* North Point, 2007.

Baldry, P. E. *The Battle Against Bacteria.* Cambridge University Press, 1965.

Barnett, JoEllen. *Time's Pendulum: The Quest to Capture Time – From Sundials to Atomic Clock.* Thomson Learning, 1999.

Bartky, I. R. 'The Adoption of Standard Time,' *Technology and Culture* 30 (1989): 48–49.

Basker, Emek. 'Raising the Barcode Scanner: Technology and Productivity in the Retail Sector,' *American Economic Journal: Applied Economics* 4, no. 3 (2012): 1–27.

Berger, Harold. *The Mystery of a New Kind of Rays: The Story of Wilhelm Conrad Roentgen and His Discovery of X-Rays.* CreateSpace Independent Publishing Platform, 2012.

Blair, B. E. 'Precision Measurement and Calibration: Frequency and Time,' *NBS Special Publication* 30, no. 5, selected NBS Papers on Frequency and Time.

Blum, Andrew. *Tubes: A Journey to the Center of the Internet.* Ecco, 2013.

Brown, George P. *Drainage Channel and Waterway: A History of the Effort to Secure an Effective and Harmless Method for the Disposal of the Sewage of the City of Chicago, and to Create a Navigable Channel Between Lake Michigan and the Mississippi River.* General Books, 2012.

Brown, Leonard. *John Coltrane and Black America's Quest for Freedom: Spirituality and the Music.* Oxford University Press, 2010.

Bruck, Hermann Alexander. *The Peripatetic Astronomer: The Life of Charles Piazzi Smyth.* Taylor & Francis, 1988.

Burian, S. J., Nix, S. J., Pitt, R. E., and Durrans, R. S. 'Urban Wastewater Management in the United States: Past, Present, and Future,' *Journal of Urban Technology* 7, no. 3 (2000): 33–62.

Cain, Louis P. 'Raising and Watering a City: Ellis Sylvester Chesbrough and Chicago's First Sanitation System,' *Technology and Culture* 13, no. 3 (1972): 353–372.

Chesbrough, E. S. 'The Drainage and Sewerage of Chicago,' paper read (explanatory and descriptive of maps and diagrams) at the annual meeting in Chicago, September 25, 1887.

Clark, G. 'Factory Discipline,' *The Journal of Economic History* 54, no. 1 (1994): 128–163.

Clegg, Brian. *Roger Bacon: The First Scientist.* Constable, 2013.

The Clorox Company: 100 Years, Reasons. The Clorox Company, 2013.

Cohn, Scotti. *It Happened in Chicago.* Globe Pequot, 2009.

Courtwright, David T. *Forces of Habit: Drugs and the Making of the Modern World.* Harvard University Press, 2002.

Cutler, D., and Miller, G. 'The Role of Public Health Improvements in Health Advances: The Twentieth-Century United States,' *Demography* 42, no. 1 (2005): 1–22.

De Landa, Manuel. *War in the Age of Intelligent Machines*. Zone, 1991.

Dickens, Charles. *Hard Times*. Knopf, 1992.

Diekman, Diane. *Twentieth Century Drifter: The Life of Marty Robbins*. University of Illinois Press, 2012.

Dolin, Eric Jay. *Leviathan: The History of Whaling in America*. Norton, 2007.

Douglas, Susan J. *Inventing American Broadcasting, 1899–1922*. Johns Hopkins University Press, 1989.

Drake, Stillman. *Galileo at Work: His Scientific Biography*. Dover, 1995.

Dreiser, Theodore. 'Great Problems of Organization, III: The Chicago Packing Industry,' *Cosmopolitan* 25 (1895).

Dreyfus, John. *The Invention of Spectacles and the Advent of Printing*. Oxford University Press, 1998.

Ekirch, Roger. *At Day's Close: A History of Nighttime*. Phoenix, 2006.

Essman, Susie. *What Would Susie Say Bullsh*t Wisdom About Love, Life, and Comedy*. Simon & Schuster, 2010.

Fagen, M. D., ed. *A History of Engineering and Science in the Bell System: National Service in War and Peace (1925–1975)*. Bell Labs, 296–317.

Fang, Irving E. *A History of Mass Communication: Six Information Revolutions*. Focal, 1997.

Fisher, Leonard Everett. *The Glassmakers (Colonial Craftsmen)*. Cavendish Square Publishing, 1997.

Fishman, Charles, *The Big Thirst: The Secret Life and Turbulent Future of Water*. Free Press, 2012.

Flanders, Judith. *Consuming Passions: Leisure and Pleasures in Victorian Britain*. Harper Perennial, 2007.

Foster, Russell, and Kreitzler, Leon. *Rhythms of Life: The Biological Clocks That Control the Daily Lives of Every Living Thing*. Yale University Press, 2005.

Freeberg, Ernest. *The Age of Edison: Electric Light and the Invention of Modern America*. Penguin, 2013.

Friedel, Robert D., Israel, Paul, and Finn, Bernards S. *Edison's Electric Light: The Art of Invention*. Baltimore: Johns Hopkins University Press, 2010.

Frost, Gary L. 'Inventing Schemes and Strategies: The Making and Selling of the Fessenden Oscillator,' *Technology and Culture* 42, no. 3 (2001): 462–488.

Gertner, Jon. *The Idea Factory: Bell Labs and the Great Age of American Innovation*. Penguin, 2013.

Gladstone, J. 'John Gorrie, The Visionary. The First Century of Air Conditioning,' *The Ashrae Journal*, article 1 (1998).

Gleick, James. *Faster: The Acceleration of Just About Everything*. Vintage, 2000.

Gleick, James. *The Information: A History, a Theory, a Flood*. Vintage, 2012.

Goetz, Thomas. *The Remedy: Robert Koch, Arthur Conan Doyle, and the Quest to Cure Tuberculosis*. Penguin, 2014.

Gray, Charlotte. *Reluctant Genius: Alexander Graham Bell and the Passion for Invention*. Arcade, 2011.

Haar, Charles M. *Mastering Boston Harbor: Courts, Dolphins, and Imperiled Waters*. Harvard University Press, 2005.

Hall, L. 'Time Standardization.' http://railroad.lindahall.org/essays/time-standardization.html.

Hamlin, Christopher. *Cholera: The Biography*. Oxford University Press, 2009.

Hecht, Jeff. *Beam: The Race to Make the Laser*. Oxford University Press, 2005.

Hecht, Jeff. *Understanding Fiber Optics*. Prentice Hall, 2005.

Heilbron, John L. *Galileo*. Oxford University Press, 2012.

'Henry Ford and the Model T: A Case Study in Productivity' (Part 1). http://www.econedlink.org/lessons/index.php?lid=668&type=student

Herman, L. M., Pack, A. A., and Hoffmann-Kuhnt, M. 'Seeing Through Sound: Dolphins Perceive the Spatial Structure of Objects Through Echolocation,' *Journal of Comparative Psychology* 112 (1998): 292–305.

Hijiya, James A. *Lee DeForest and the Fatherhood of Radio*. Lehigh University Press, 1992.

Hill, Libby. *The Chicago River: A Natural and Unnatural History*. Lake Claremont Press, 2000.

Howse, Derek. *Greenwich Time and the Discovery of the Longitude*. Oxford University Press, 1980.

Irwin, Emily. 'The Spermaceti Candle and the American Whaling Industry,' *Historia* 21, 2012.

Jagger, Cedric. *The World's Greatest Clocks and Watches*. Galley Press, 1987.

Jefferson, George, and Lowell, Lindsay. *Fossil Treasures of the Anza-Borrego Desert: A Geography of Time*. Sunbelt Publications, 2006.

Jonnes, Jill. *Empires of Light: Edison, Tesla, Westinghouse, and the Race to Electrify the World*. Random House, 2004.

Klein, Stefan. *Time: A User's Guide*. Penguin, 2008.

Klooster, John W. *Icons of Invention: The Makers of the Modern World from Gutenberg to Gates*. Greenwood, 2009.

Koestler, Arthur. *The Act of Creation*. Penguin, 1990.

Kurlansky, Mark. *Birdseye: The Adventures of a Curious Man*. Broadway Books, 2012.

Landes, David S. *Revolution in Time: Clocks and the Making of the Modern World*. Belknap Press, 2000.

Livingston, Jessica. *Founders at Work: Stories of Startups' Early Days*. Apress, 2008.

Lovell, D. J. *Optical Anecdotes*. SPIE Publications, 2004.

Macfarlane, Alan, and Martin, Gerry. *Glass: A World History*. University of Chicago Press, 2002.

Macrae, David. *The Americans at Home: Pen-and-ink Sketches of American Men, Manners and Institutions, Volume 2*. Edmonston & Douglas, 1870.

Maier, Pauline. *Inventing America: A History of the United States, Volume 2*. Norton, 2005.

Matthew, Michael R., Clough, Michael P., and Ogilvie, C. 'Pendulum Motion: The Value of Idealization in Science.' http://www.storybehindthescience.org/pdf/pendulum.pdf.

McCrossen, Alexis. *Marking Modern Times: A History of Clocks, Watches, and Other Timekeepers in American Life*. University of Chicago Press, 2013.

McGuire, Michael J. *The Chlorine Revolution*. American Water Works Association, 2013.

Mercer, David. *The Telephone: The Life Story of a Technology*. Greenwood, 2006.

Millard, Andre. *America on Record: A History of Recorded Sound*. Cambridge University Press, 2005.

Miller, Donald L. *City of the Century: The Epic of Chicago and the Making of America*. Simon & Schuster, 1996.

Morris, Robert D. *The Blue Death: Disease, Disaster, and the Water We Drink*. Harper, 2007.

Mumford, Lewis. *Technics and Civilisation*. Routledge, 1934.

Ness, Roberta. *Genius Unmasked*. Oxford University Press, 2013.

Ngozika Ihewulezi, Cajetan. *The History of Poverty in a Rich and Blessed America: A Comparative Look on How the Euro-Ethnic Immi-*

grant Groups and the Racial Minorities Have Experienced and Struggled Against Poverty in American History. Authorhouse, 2008.

Nicolson, Malcolm, and Fleming, John E. E. 'Imaging and Imagining the Foetus: The Development of Obstetric Ultrasound.' John Hopkins University Press, 2013.

Ollerton, J., and Coulthard, E. 'Evolution of Animal Pollination,' *Science* 326.5954 (2009): 808–809.

Pack, A. A., and Herman, L. M. 'Sensory Integration in the Bottlenosed Dolphin: Immediate Recognition of Complex Shapes Across the Senses of Echolocation and Vision,' *Journal of the Acoustical Society of America* 98 (1995): 722–733.

Pack, A. A., Herman, L. M., and Hoffmann-Kuhnt, M. 'Dolphin echolocation shape perception: From Sound to Object.' In J. Thomas, C. Moss, and Vater, M. (eds.).

Pack, A. A., and Herman, L. M. 'Seeing Through Sound: Dolphins (Tursiops truncatus) Perceive the Spatial Structure of Objects Through Echolocation,' *Journal of Comparative Psychology* 112, no. 3 (1998): 292–305.

Pascal, Janet B. *Jacob Riis: Reporter and Reformer*. Oxford University Press, 2005.

Pendergrast, Mark. *Mirror Mirror: A History of the Human Love Affair with Reflection*. Basic Books, 2004.

Clair C. Patterson (1922–1995), interviewed by Shirley K. Cohen. March 5, 6, and 9, 1995. Archives California Institute of Technology, Pasadena, California. http://oralhistories.library.caltech.edu/32/1/OH_Patterson.pdf.

Poe, Marshall T. *A History of Communications: Media and Society from the Evolution of Speech to the Internet*. Cambridge University Press, 2010.

Polsby, Nelson W. *How Congress Evolves: Social Bases of Institutional Change*. Oxford University Press, 2005.

Praeger, Dave. *Poop Culture: How America Is Shaped by Its Grossest National Product*. Feral House, 2007.

Price, R. 'Origins of the Waltham Model 57.' Copyright © 1997–2012 Price-Less Ads http://www.pricelessads.com/m57/monograph/main.pdf

Priestley, Philip T. *Aaron Lufkin Dennison – an Industrial Pioneer and His Legacy*. National Association of Watch & Clock Collectors, 2010.

Ranford, J. L. *Analogue Day*. Ranford, 2014.

Rhodes, Richard. *Hedy's Folly: The Life and Breakthrough Inventions of Hedy Lamarr, the Most Beautiful Woman in the World*. Vintage, 2012.

Ribbat, Christoph. *Flickering Light: A History of Neon*. Reaktion Books, 2013.

Richards, E. G. *Mapping Time: The Calendar and Its History*. Oxford University Press, 2000.

Riis, Jacob A. *How the Other Half Lives: Studies among the Tenements of New York*. Dover, 1971.

Roberts, Sam. *Grand Central Station: How a Station Transformed America*. Grand Central Publishing, 2013.

Royte, Elizabeth. *Bottlemania: How Water Went On Sale And Why We Bought It*. Bloomsbury, 2008.

Shachtman, Tom. *Absolute Zero and the Conquest of Cold*. Houghton Mifflin, 1999.

Schlesinger, Henry. *The Battery: How Portable Power Sparked a Technological Revolution*. Harper Perennial, 2011.

Schwartz, Hillel. *Making Noise – From Babel to the Big Bang and Beyond*. MIT Press, 2011.

Senior, John E. *Marie and Pierre Curie*. Sutton Publishing, 1998.

Silverman, Kenneth. *Lightning Man: The Accursed Life of Samuel F. B. Morse*. Da Capo Press, 2004.

Sinclair, Upton. *The Jungle*. Dover, 2001.

Skrabec, Quentin R., Jr. *Edward Drummond Libbery: American Glassmaker*. McFarland, 2011.

Sterne, Jonathan. *The Audible Past: Cultural Origins of Sound Reproduction*. Duke University Press, 2003.

Steven-Boniecki, Dwight. *Live TV: From the Moon*. Apogee Books, 2010.

Stross, Randall E. *The Wizard of Menlo Park: How Thomas Alva Edison Invented the Modern World*. Crown, 2007.

Swade, Doron. *The Difference Engine: Charles Babbage and the Quest to Build the First Computer*. Penguin, 2002.

Taylor, Nick. *Laser: The Inventor, the Nobel Laureate, and the Thirty-Year Patent War*. Backprint.com, 2007.

Thompson, Emily. *The Soundscape of Modernity: Architectural Acoustics and the Culture of Listening in America, 1900–1933*. MIT Press, 2004.

Thompson, E. P. 'Time, Work-Discipline and Industrial Capitalism,' *Past & Present* 38 (1967): 56–97.

Thoreau, Henry David. *Walden*. Phoenix, 1995.

Toole, Betty Alexandra. *Ada, the Enchantress of Numbers: Poetical Science*. Critical Connection, 2010.

Toso, Gianfranco. *Murano Glass: A History of Glass*. Arsenale, 1999.

Venturi, Robert, Scott Brown, Denise, and Izenour, Steven. *Learning From Las Vegas*. MIT Press, 1977.

Veritá, Marco. 'L'invenzione del cristallo muranese: Una verifica ana litica delle fonti storiche,' *Rivista della Stazione Sperimental del Vetro* 15 (1985): 17–29.

Watson, Peter. *Ideas: A History: From Fire to Freud*. Phoenix, 2006.

Weightman, Gavin. *The Frozen Water Trade: How Ice from New England Kept the World Cool*. HarperCollins, 2003.

Wells, H. G. *The War of the Worlds*. New American Library, 1986.

Wheen, Andrew. *Dot-Dash to Dot.Com: How Modern Telecommunications Evolved from the Telegraph to the Internet.* Springer, 2011.

White, M. 'The Economics of Time Zones,' March 2005. www.learningace.com/doc/1852927/fbfb4e95bef9efa4666d23729d3aa5b6/timezones

Willach, Rolf. *The Long Route to the Invention of the Telescope.* American Philosophical Society, 2008.

Wilson, Bee. *Swindled: The Dark History of Food Fraud, from Poisoned Candy to Counterfeit Coffee.* Princeton University Press, 2008.

Wiltse, Jeff. *Contested Waters: A Social History of Swimming Pools in America.* University of North Carolina Press, 2010.

Wolfe, Tom. *The Kandy-Kolored Tangerine-Flake Streamline Baby.* Picador, 2009.

Woods-Marsden, Joanna. *Renaissance Self-Portraiture: The Visual Construction of Identity and the Social Status of the Artist.* Yale University Press, 1998.

Woolley, Benjamin. *The Bride of Science: Romance, Reason, and Byron's Daughter.* McGraw-Hill, 2000.

Wright, Lawrence. *Clean and Decent: The Fascinating History of the Bathroom and the Water Closet.* Routledge & Kegan Paul, 1984.

Yochelson, Bonnie. *Rediscovering Jacob Riis: The Reformer, His Journalism, and His Photographs.* New Press, 2008.

Yong, Ed. 'Hummingbird Flight Has a Clever Twist,' *Nature* (2011).

Zeng, Yi, et al. 'Causes and Implications of the Recent Increase in the Reported Sex Ratio at Birth in China,' *Population and Development Review* 19, no. 2 (1993): 294–295.

Index